SpringerBriefs in Cyb

Cybersecurity is a difficult and complex field. The technical, political and legal questions surrounding it are complicated, often stretching a spectrum of diverse technologies, varying legal bodies, different political ideas and responsibilities. Cybersecurity is intrinsically interdisciplinary, and most activities in one field immediately affect the others. Technologies and techniques, strategies and tactics, motives and ideologies, rules and laws, institutions and industries, power and money—all of these topics have a role to play in cybersecurity, and all of these are tightly interwoven.

The *SpringerBriefs in Cybersecurity* series is comprised of two types of briefs: topic- and country-specific briefs. Topic-specific briefs strive to provide a comprehensive coverage of the whole range of topics surrounding cybersecurity, combining whenever possible legal, ethical, social, political and technical issues. Authors with diverse backgrounds explain their motivation, their mindset, and their approach to the topic, to illuminate its theoretical foundations, the practical nuts and bolts and its past, present and future. Country-specific briefs cover national perceptions and strategies, with officials and national authorities explaining the background, the leading thoughts and interests behind the official statements, to foster a more informed international dialogue.

More information about this series at http://www.springer.com/series/10634

Greg Austin

Cybersecurity in China

The Next Wave

 Springer

Greg Austin
Australian Centre for Cyber Security
University of New South Wales
Canberra, ACT
Australia

ISSN 2193-973X   ISSN 2193-9748   (electronic)
SpringerBriefs in Cybersecurity
ISBN 978-3-319-68435-2   ISBN 978-3-319-68436-9   (eBook)
https://doi.org/10.1007/978-3-319-68436-9

Library of Congress Control Number: 2018939006

Printed on acid-free paper

This Springer imprint is published by the registered company Springer International Publishing AG
part of Springer Nature
The registered company address is: Gewerbestrasse 11, 6330 Cham, Switzerland

Foreword

China is certainly among the most interesting players in cybersecurity. While oftentimes rather visible as intrusive and offensive, for instance, through its massive efforts to conduct mass surveillance of its own citizens or in its (alleged) global cyber industrial espionage campaigns, the country is also in a rather unique position regarding its defensive cybersecurity outset and interests. It is, by now, heavily industrialized, fully connected, with a large number of foreign companies, bringing their own technologies with them, and its citizens are increasingly more digitalized and connected through the Internet. Accordingly, the country is highly and increasingly vulnerable to cyberattacks or information operations, and strong concerns about its inner and outer security ensue. So strong, in fact, that cybersecurity has been made a senior management priority, with Secretary General Xi Jinping seeing to those matters personally. China in fact considers its cybersecurity a key matter of its sovereignty and keeps stressing its right and duty to create "cyber sovereignty" in many aspects.

The outcome is in many cases very interesting and individual—and of high impact internationally. The close entanglement of technical security, censorship, and surveillance is one example. Another more recent one is the new regulation of foreign companies operating in China. Foreign companies now have to use Chinese security technologies and encryption, encrypted connections are switched off, and even the source code produced or used in China has to be disclosed to the Chinese government. As a result, all these companies will turn more and more transparent. Given China's interest in industrial espionage, these efforts have been a cause of (mostly silent) outrage abroad and will have wider strategic implications for the Chinese economy—but may also provide a model for other countries striving to achieve cyber sovereignty.

In sum, China is and will continue to be a very interesting place to observe when it comes to the regulation of cybersecurity. Accordingly, it is a great benefit for this SpringerBrief in cybersecurity to have this particular brief about cybersecurity in China, and an even greater benefit to have it written by Greg Austin. Greg is in close touch with China and its cybersecurity regulators for many years, theoretically and practically, and can not only provide a full and credible account of the country's efforts, but also background stories and additional knowledge which can not be found anywhere else. Founded in his competence, this SpringerBrief will help to understand China and its efforts and provide important insights and details for academics, diplomats, and industries alike.

Berlin, Germany   Dr. Sandro Gaycken
February 2018

Preface

China has established a global reputation for cyberattack. How good is it at cyber defense? This book offers a health check, a report card, on China's cybersecurity system in the face of escalating threats from criminal gangs at home and abroad, as well as from hostile states. The book hopes to contribute some new foundations for a comprehensive benchmarking of China's responses to the problem of security in cyberspace. But it cannot claim to offer a comprehensive analysis. The scope of the subject of cybersecurity is simply too broad. In acknowledging that, the book hopes to inform current approaches to assessment of national cybersecurity performance to stimulate more granular research across the broader set of issues that are canvassed in this book.

At a foundational level, the Bell Labs model of cybersecurity distinguishes between eight ingredients of the problem set: software, hardware, networks, payload, power (electricity supply), people, ecosystem, and policy (Rauscher et al. 2006). While each of these may appear at first glance to be relatively compact and bounded for any country, they are individually hugely complex, and have become more so, as technologies like mobile computing, quantum computing, cloud computing, the internet of things (IoT), and advanced artificial intelligence have come into play. Even as understood within the eight-ingredient framework from the engineer's perspective, the field of cybersecurity is highly dynamic.

In defining the boundaries of the subject of cybersecurity according to the eight ingredients, we also need to be cognizant of the different practices, technologies, personnel, and organizations needed for distinct mission sets, such as countering cybercrime against corporations or citizens, online child protection, preventing political subversion, countering cyber espionage, protecting critical infrastructure, and preparing for cyber-enabled warfare.

The subject in this case is China in the broad, but we need to break that down into at least three stakeholder subsets: its government, its corporations, and its citizens. Boundaries between the three sets of subjects in cyber policy and activities are in many circumstances quite fluid. These three sets of stakeholder have many different interests, face many different threats, and have variegated response capabilities.

The three stakeholder sets (each with many subsets) represent fundamentally different social and political phenomena. We must avoid accepting at face value the Chinese government's view that it speaks for cybersecurity in China. It wants to, but as this book demonstrates, it is in the character of the information age that a government cannot be the source of cybersecurity for its corporations or citizens. If we measure government capacity, we are only measuring cybersecurity for the government, not for the country. The Chinese government blurs this distinction, unlike governments in the UK and the USA. These liberal democracies are firmly of the view that the government is not responsible for the cybersecurity of its corporations and citizens when they confront a threat. These two governments limit themselves to providing assistance as they can to non-government actors, both in advance of or during an attack. In most countries, governments have inadequate resources to provide cybersecurity for their own agencies, let alone their corporations and citizens. This stakeholder distinction is becoming more important in China since the central government is proving to be far less capable in delivering cybersecurity for the whole country than it would like to be.

There is a fourth category of stakeholders in Chinese cyberspace: foreign governments, foreign corporations, and foreign citizens. Cybersecurity in the country is an unavoidably international and globalized activity. While this book concentrates on the domestic scene in China, it must also sketch broad outlines of the interaction between Chinese and non-national influences as they take place inside the country.

Thus, the picture presented in the book of cybersecurity in China is a mosaic or a multi-dimensional montage. Conflict between and within the different sets of stakeholders and cybersecurity mission sets in China, as in other countries, is assumed. That said, evidence of the contours of those conflicts in China is often sparse as a result of the country's non-transparent political system.

The contest between and among the different interest groups and mission sets analyzed in this book speaks to the centrality of the concept of power as the ultimate test of national cybersecurity. Many extant measures of cybersecurity at the national level focus on inputs (announced government measures) rather than on results or outcomes (the manifestations of cyber power). This book proposes a new orientation in assessing national cybersecurity: one that relates to the actual exercise of power in cyberspace (the outcomes) in the pursuit of security. In simple terms, the quality of cybersecurity in any country will be defined in larger part by degrees of implementation of policies by a variety of stakeholders, rather than being the sum of policy declarations and intents.

This book is a sequel to *Cyber Policy in China* (Austin 2014), which takes a values-based approach in analyzing China's ambitions of becoming an advanced information society, reviewing political, economic and security aspects, on both the domestic and international fronts. It referenced the many excellent books on cyber censorship and political activism in cyberspace in China that had been written up that time. It also referenced key works on the political economy of China's ICT sector. That book included analysis of the landmark statement by President Xi Jinping in February 2014 that China would do everything needed for the country to become a cyber power. The current book, *Cybersecurity in China*, has a narrower focus on two fronts. First, it has a sharper domestic focus, leaving international considerations somewhat to one side, though not ignoring them completely. Secondly, it focuses exclusively on China's security in cyberspace not its wider "information society" ambition.

This book relies on China's current official definition of cybersecurity, discussed in more detail in Chapter 1. This definition is essentially the same as we might find in other countries—in spite of linguistic preferences around certain renderings of the concept in Chinese. These include: *wǎngluò ānquán*, translated variously by Chinese authors into English as "network security," "cybersecurity," or "cyberspace security"; and *xìnxī ānquán* which is a cognate for the English term "information security." The Chinese approach sees cybersecurity (and/or information security) as a subset of national security. This means that cybersecurity is, for China, as other leading powers, a socio-technical phenomenon: a process or state of mind that seeks to minimize actual and perceived risks to a subject's well-being arising from activities in cyberspace (Austin 2017).

In Chinese scholarship, the socio-technical phenomenon of cybersecurity is a relatively new field of study that takes place in a highly distorted information ecosystem. Universities in the country have been slow to take up the socio-technical dimensions of cybersecurity, and few have dared to undertake critical analysis of politically sensitive aspects of cybersecurity (the larger share of the subject). Leading Chinese studies in the public domain as do exist have been reviewed in preparation of this book.

In scholarship outside China, there are several useful descriptive book-length studies on cybersecurity policy in China published in recent years (ICGC 2012; Ventre 2014; Lindsay et al. 2015; Inkster 2016; Raud 2016; Cheng 2017). In shorter formats, scholars such as Yang Guobin, Adam Segal at the Council on Foreign Relations, Rogier Creemers at Oxford, Nir Kshetri, a group at the Jamestown Foundation (especially Peter Mattis), Joe McReynolds and Leigh Ann Ragland at the Defense Group Inc., Russell Hsiao at Project 2049, the China Media Project at Hong Kong University, and the team at the China Digital Times have regularly made insightful contributions to the body of knowledge on different aspects of China's most recent cybersecurity policies. Similarly strong contributions have been made in a variety of online formats by numerous professionals in foreign ICT corporations and law firms working in or with China. As a result, we know quite well the system for managing the diverse aspects of cybersecurity in China, and the chronological sequence of announced policy responses.

Kshetri (2013: 1) has characterized the general situation well: the "distinctive pattern of the country's cyberattack and cybersecurity landscapes" is explained by "China's global ambition, the shift in the base of regime legitimacy from Marxism Leninism to economic growth, [and] the strong state and weak civil society." One report (Jamestown 2014) observed that "research on Chinese cyber defense lags behind assessment of China's offensive capabilities" and concluded that the resulting approach is "fragmented both by region and function that is further complicated by turf battles between regulators institutions." This book adds contours and depth to such broad assessments offered in shorter formats.

The book also offers an assessment of the effectiveness of efforts in China to protect various interests in cyberspace. In this analysis, the views of Chinese scholars, officials, and other opinion leaders are essential, but through the research for this book, the author formed the view that many key Chinese sources have been ignored in Western scholarship. For a number of reasons, the open-source scholarly study in China of domestic aspects of security in cyberspace as a socio-technical phenomenon is still in its infancy. It has been impossible to verify independently much of the data and many of the claims cited in this book about progress made. At the same time, as some of the material collated in this book suggests, this field of study in China is still substantial enough to be a major source of information and assessment. Moreover, there is a substantial body of knowledge on cybersecurity in China held by market research firms, such as IDC and Gartner, and most scholars do not enjoy easy access to their collections for public citation purposes.

This is the first book on cybersecurity in China to base itself around an assessment of China's cyber industrial complex.

The sequence of chapters is as follows. Chapter 1 provides an introductory snapshot of the cybersecurity ecosystem in China, setting up deeper analysis of subjects raised in later chapters. Chapter 2 provides a review of the state of university-based cybersecurity education in China. Chapter 3 offers a summary of Chinese views of the cyber industrial complex. Chapters 4–6 look in more depth at the actual state of cybersecurity for the three subsets of China called out above: the corporate sector, the citizenry, and the government. Chapter 7 offers an evaluation of the state of cybersecurity in China. A brief Chapter 8, called "The Next Wave," suggests (unsurprisingly) that China's cybersecurity in twenty years' time will look radically different from how it appears today as a result of the policy turns beginning around 2011 and boosted substantially beginning in 2014. But China does not determine its cybersecurity condition alone. This will remain a multi-factor, multi-national enterprise that has to be managed. Given China's choices to oppose the value systems of the US-led liberal democratic alliance, the scope for equanimity for China's leaders in managing cyberspace policy will remain about the same as if they were riding a tiger.

The author would like to thank colleagues in the field, especially from China, who have been open in their discussion of the subjects in this book. The author would also like to acknowledge the support on this project of several part-time research assistants (Lu Wenze, Meng Fei, and Zu Haoyue). The author alone is responsible for analysis and commentary on the material from Chinese language sources provided by the able research assistants, with the several exceptions noted and referenced.

The author is also grateful for permission from The Globalist, the best-informed and most thoughtful daily on the big issues of global affairs, and from The Diplomat, the premier Web site for analysis on Asia-Pacific security affairs, to use his material previously published by them.

Canberra, Australia   Greg Austin

References

Austin G (2014) Cyber policy in China. Polity, Cambridge

Austin G (2017) Restraint and governance in cyberspace. In: Burke A, Parker R (eds) Global insecurity: futures of chaos and governance. Palgrave

Cheng D (2017) Cyber dragon. Inside China's information warfare and cyber operations. Praeger, Santa Barbara

ICGC (2012) China and cybersecurity: political, economic, and strategic dimensions. Report from Workshops held at the University of California, San Diego April 2012

Inkster N (2016) China's cyber power. IISS, London

Kshetri N (2013) Cybercrime and cyber-security issues associated with China: some economic and institutional considerations. Electron Commer Res 13:41. https://doi.org/10.1007/s10660-013-9105-4

Lindsay JR, Cheung TM, Reveron DS (eds) (2015) China and cybersecurity: espionage, strategy, and politics in the digital domain. Oxford University Press, Oxford

Raud M (2016) China and cyber: strategies, attitudes and organization. NATO CCDCOE, Tallinn

Rauscher KF, Krock RE, Runyon JR (2006) Eight ingredients of communications infrastructure: A systematic and comprehensive framework for enhancing network reliability and security. Bell Labs Technical Journal 11.3 (2006): 73-81

Ventre D (ed) (2014) Chinese cybersecurity and defense. Wiley

Contents

Acronyms

2FA	Two-Factor Authentication
APT	Advanced Persistent Threat
CAC	Cyberspace Administration of China
CAICT	China Academy of Information and Communications Technology
CAS	Chinese Academy of Sciences
CCP	Chinese Communist Party
CCTV	China Central Television
CEC	China Electronics Information Industry Group (known in English by the acronym CEC for an earlier incarnation, China Electronics Corporation)
CERT	Computer Emergency Response Team
CESI	China Electronic Standards Institute
CETC	China Electronics Technology Group Corporation
CII	Critical Information Infrastructure
CLG	Central Leading Group
CPLC	Central Political and Legal Commission
CNCA	China National Certification Administration
CNCERT/CC	China Computer Emergency Response Team/Coordination Centre
CNITSEC	China National Information Technology Security Evaluation Centre
CNKI	China Knowledge Resource Integrated Database
CNNVD	China National Vulnerability Database for Information Security
CUAA	Chinese Universities Alumni Association
DDOS	Distributed Denial-of-Service
ICT	Information and Communications Technology
ICAO	International Civil Aviation Organization
IEEE	Institute of Electrical and Electronics Engineers
ISOC	Internet Society of China
IT	Information Technology
ITU	International Telecommunications Union

LMS	Lab for Media Search
MIIT	Ministry of Industry and Information Technology
MoE	Ministry of Education
MLPS	Multi-Level Protection System
MPS	Ministry of Public Security
MSS	Ministry of State Security
NICE	National Initiative for Cybersecurity Education
NISSTC	National Information Security Standardization Technical Committee
NPC	National People's Congress
NPU	National Police University
NSA	National Security Agency
OS	Operating System
PLA	People's Liberation Army
PSU	Public Security University
SAR	Special Administrative Region
SCO	Shanghai Cooperation Organization
SEA	State Encryption Administration
SKLOIS	State Key Laboratory of Information Security
SSB	State Secrecy Bureau
TC260	Technical Committee 260
VPN	Virtual Private Network
XJYU	Xian Jiaotong University

List of Tables

List of Boxes

Chapter 1
The Cybersecurity Ecosystem

Abstract China conducts more cyber espionage on itself than any other country does. The ecosystem for security in cyberspace is distorted by the country's political system. The chapter lays out China's definition of the problem set, and describes key features of the 2016 National Cyber Security Strategy. It then gives some background on the national policy shift in 2014 represented by Xi's declaration of the ambition for China to become a cyber power.

1.1 Introduction: Weak and Strong

China's political leaders feel insecure in cyberspace. Its corporate leaders lament the absence of a strong domestic cybersecurity industry. Its citizens are engaged in high risk online behaviors for political or personal reasons, and often out of ignorance of the dangers. Foreign corporations dominated the delivery of cybersecurity in China until quite recently. Some foreign governments work hard to undermine it. The People's Liberation Army (PLA) is struggling to come to terms with cyber war concepts and technologies beyond cyber espionage. On the other hand, the internal security agencies are among the world leaders in domestic cyber surveillance, often relying on the services and equipment of foreign corporations. China appears to do better than almost any other country in catching its own cyber criminals. Yet it has an almost invisible, probably non-existent capability for protecting national critical infrastructure in cyberspace. The country's domestic scientific base for security in cyberspace is in its infancy, and this judgment includes the scientific study of the country's overall cybersecurity. This is the multi-tiered and developmentally-challenged reality of cybersecurity in China.

Cyberspace is an inherently insecure environment. Even the United States, the world's wealthiest and most technologically advanced country, faces chronic cyber insecurities. But three factors combine to make Chinese users particularly insecure compared with those of other G20 countries, including India, a similarly large developing country. These three factors are: the numerically large number of users (including both humans and other machines, the former with little of the necessary security

© The Author(s) 2018
G. Austin, *Cybersecurity in China*, SpringerBriefs in Cybersecurity,
https://doi.org/10.1007/978-3-319-68436-9_1

literacy), the higher prevalence of pirated software, and a very high intensity of consumption of ICT products (a "rush to use").

At a deeper level of analysis, but one which has high cogency for daily security in cyberspace in China is what we might call the "spy versus spy" effect. China is the only country which is targeted every second across a wide spectrum of social and economic life by two of the most powerful cyber actors in the world. One of these two perpetrators is obvious. The United States has built and operates the *largest cyberspace military and espionage alliance in human history* and one of its primary targets is China—the country as whole, not just the government. The second perpetrator is less obvious. It is the government of China. It spies on itself. It has built and operates the *largest cyber-enabled internal security surveillance system in human history* and the primary target of this system is the country as a whole, including the government and its 80 million CCP members. Within two to three decades, advances in artificial intelligence may allow China to achieve a near-perfect replica of the Orwellian vision of Big Brother's omniscient domestic social surveillance seen in the novel *1984*.

If you are an adult inside China, regardless of your nationality, occupation or location, no matter what cyber system you operate and depend on, there is a very high chance that it is being surveilled or can be surveilled on short notice by both of the two cyber superpowers. There is also a high chance that they will have developed an option for attack on your cyber lifelines and foundations. To undertake this surveillance, the two countries have devised access technologies for the cyber systems you use. This is not the case for any other country in the world apart from China. While such surveillance and the cyber insecurity needed for it (endless intrusions) would be illegal in some countries, neither the U.S. government or Chinese government feels constrained by law in this activity inside China. The operating norms (one in international law and one in domestic law) are that cyber attacks for espionage purposes in this case are lawful, whether it U.S. espionage against China or Chinese political surveillance of its own citizens.

The weakness of China's cybersecurity is one of the best kept secrets in Washington because of the latter's interest in exploiting the vulnerability for intelligence and military purposes. For its part, the Chinese leadership faces competing priorities. On the one hand, China could reveal its own weaknesses fully in order to promote higher levels of awareness and begin to structure adequate responses. On the other hand, to do so would undercut both the political legitimacy of the government as defender of the country's interests and undermine the government's ability to exploit the weaknesses for its own internal political intelligence collection. In fact, the country's leadership has a vested interest in convincing the Chinese public that it is omnipotent in cyberspace and that it can, in the fullness of time, protect CCP interests in the infosphere.

There is another general condition that undermines security in cyberspace that is worthy of note. Companies that provide services in this field in one country become a prime target for its intelligence adversaries, a fact recognized at least in the case of the U.S. agencies (Bing 2017). The reason is that the companies "have direct, continuous access into their clients' networks and collect large quantities of data about them".

1.2 Cyber Insecurity: Snapshot

By way of introduction to the story of cybersecurity in China, here are five reports of activity in January 2017 that capture many of the challenges faced by stakeholders in the country, both at the time and on a continuing basis. Some aspects of these stories are unique to China, while others are not.

Personal Cyber Dependency and Government Intervention: A Chinese agency announced a draft regulation that would require the providers of online games to enforce a curfew (from midnight to 8.00 am) in a bid to restrict "addictive" use by minors (China Daily 2017a).

Legal Obligations of Service Providers: The Guangdong Province branch of the Cyberspace Administration of China (CAC) reported that it had taken down over 5000 illegal apps, of which more than 1200 had carried pornography (Xinhua 2017a). A number of Chinese corporations were named with the implication that they were not being vigilant enough (Tencent, China Mobile, Huawei, ZTE, Coolpad, Meizu, Oppo and Vivo).

Cyber Crime Statistics: With the start of the new year in 2017, cyber crime fighters reported the scale of the problem over the previous year, citing the arrest of 19,345 suspects of cyber or telecommunications fraud throughout China (Xinhua 2017b). This was small share of total cyber crime in China, with Guangdong province alone reporting its 2016 situation in the following terms: "4125 cyber crimes and ... 15,000 criminal suspects from 892 criminal gangs" (Mi 2017). Guangdong, adjacent to Hong Kong, is one of the richest and most connected provinces in the country. (China has 31 provinces, including municipalities and autonomous regions with similar status as provinces. It also has two Special Administrative Regions or SARs, Hong Kong and Macao.)

Censorship and Leadership Security: A monitoring organization based in Hong Kong, China Digital Times (2017), reported that the relevant Ministry had ordered a crackdown on virtual private networks (VPNs) which users rely on to access international internet content that is banned or blocked by the government. Chinese regulators were moving to prohibit the setting up or renting of a communication channel, such as a VPN, to undertake cross-border communications unless the channel is approved by the government. One of the primary reasons why the authorities are so keen to shut down this access is to prevent the spreading of stories about the wealth and alleged corruption of political leaders which have been reported on several times in detail by foreign media since 2012.

U.S. Military Superiority in Cyberspace: The chinamil.com website carried a story from the *People's Liberation Army Daily*, which reported cyber warfare capability as one of the main success stories of the Obama Administration (Feng 2017). The report left unspoken what the implications for China of this evolution might be, but it replays subtly the long-standing Chinese view that the country lags badly behind the United States in cyber military power because of what the article calls that country's "IT advantages".

1.3 Toward a More Secure Cyberspace: Snapshot

To complement the above snapshot of cyber insecurity in the month of January 2017, here is a similar snapshot of events in the following month suggestive of enhanced cybersecurity in China.

Cybersecurity Review Committee: One of the more significant developments was the announcement on 8 February 2017 that the government would set up a cybersecurity review committee with a wide-ranging supervisory role for products and services affecting national security or public interest ("to deliberate important policies on cybersecurity and organize reviews") (China Daily 2017b). In releasing the news, the CAC said other issues for review would include risks of criminal use to "illegally gather, store, process or make use of" consumer information and unfair competition practices. The result would be blacklisting of services or products that do not pass the review, resulting in a ban on their use by government agencies, the CCP, or strategic industries. CAC promised that the review process would treat foreign and domestic suppliers equally, touching on a major concern of the former that the country is discriminating in favour of domestic companies.

Fencing China Off: On 17 February, President Xi Jinping used a seminar convened under the auspices of the National Security Commission, also then a relatively new body which he chairs, to call for attention to cybersecurity as part of a new "global perspective" on protection of the country. Advocating greater attention to prevention of risks, Xi promoted consolidation of the cybersecurity "fence" and "improved efforts to secure information infrastructure" (China.org 2017).

Domestic Cyber Industry: On the technology front, as a sign of things to come, a Chinese-owned company based in Hong Kong, Nexusguard, took advantage of the release of the latest Top 500 global ranking of cybersecurity companies in February 2017 to take pride in its consistent placement in the top 25, ahead even of Russia's better known Kaspersky Lab (AsiaOne 2017). Nexusguard, which specialises in protection against distributed denial of service (DDOS) attacks, is wholly owned by Legend Venture Holdings, which is the largest shareholder in Lenovo, the Chinese company that became the world's largest manufacturer of desktop computers after buying out IBM's PC business.

1.4 China's Definition of Cybersecurity

China as a whole has come late to a comprehensive view of cybersecurity as a socio-technical phenomenon. A useful reference point is the three stage distinction devised by Hathaway and Klimburg (2012: 29–31), that is between a whole-of-government approach, to a whole-of-nation approach, to a whole-of-systems. China flagged its intention to pursue a whole-of-nation approach in its first regulations on cybersecurity in Decree No. 147 of the State Council on 18 February 1994 (State Council 1994). The regulations contained therein were rudimentary (expressed in some places in terms of computer worms or viruses), and could largely have been considered enterprise-level methodologies. Over the subsequent decade, the government eventually became much more focused and began to frame up over-arching concepts, producing in 2007, as Marro (2016) observed, a five-tier classification system for impacts on information security developed by the Ministry of Public Security (MPS) (www.gov.cn 2007).

This system, known by the English rubric MLPS, applies nationality and corporate governance tests to firms operating at level three or higher in this classification system:

1. damage to the legitimate rights and interests of citizens, legal persons and other organizations, without prejudice to national security, social order and public interest
2. serious damage to the legitimate rights and interests of citizens, legal persons and other organizations, or damage to the social order and public interests, without prejudice to national security
3. it will cause serious damage to social order and public interest, or damage to national security
4. particularly serious damage to social order and public interest, or cause serious damage to national security
5. particularly serious damage to national security.

The regulations also allow for MPS supervision of any information activity inside China regardless of nationality if level 2, or higher, of the five-tier system is in play. The regulations foreshadow a comprehensive approach but for most of the time since the 1994 regulations, China's agencies and industries involved in cyber security had been operating under a whole-of-nation approach.

By 2016, China moved to a very modern concept of cybersecurity that conformed to the whole-of-system approach. It was laid out in a speech by Xi (2016) to the National Meeting on Cybersecurity and Informatization. The key points of this concept have been reproduced in Box 1.1. It should be noted that if 2016 is the departure date for articulation of a whole-of-system approach, then effective implementation in many areas of policy will be a process that takes decades from now rather than a few years.

Box 1.1: Xi Jinping on the Character of Cybersecurity

1. cybersecurity is holistic rather than fragmented. In the information age, cybersecurity has a close relationship with many other aspects of national security.
2. cybersecurity is dynamic rather than static. Information technology changes faster and faster, and the networks, which used to be scattered and independent, become highly connected and interdependent. The threat sources of cybersecurity and the means of attack are constantly developing. The idea of relying on a few pieces of security equipment and security software to keep safety is outdated. It needs to establish a dynamic, integrated protection concept.
3. cybersecurity is open rather than closed. The cybersecurity level can continue to improve only if we strengthen international exchange, cooperation, and interaction to absorb advanced technology.
4. cybersecurity is relative rather than absolute. There is no absolute security, we should consider basic national conditions and avoid the pursuit of absolute security regardless of cost.
5. cybersecurity is common rather than isolated. Cybersecurity is for people, and relies on people. Cybersecurity is the responsibility of the whole society, and it needs the joint participation of the government, enterprises, social organizations, and most Internet users to build a line of defence.

Xi not only charted a general direction for the development of China's cybersecurity industry but also emphasized the importance of dealing correctly with the relationship between security and development. He called for acceleration of the construction of security for critical information infrastructure (CII), especially in the fields of finance, energy, electricity and communication, describing it as the "nerve centre of economic and social operation". Xi observed that China's CII was the "top priority of cybersecurity" since it was already the "target of serious attack".

Xi called for greater recognition of monitoring cybersecurity "comprehensively and around the clock", and for the country to develop a "unified and efficient risk reporting mechanism for cybersecurity, an information sharing mechanism, and a disposal [dissemination] mechanism to accurately grasp the rules and trends of cybersecurity". He also called for enhanced defensive ability and deterrence in cyberspace: "The essence of cybersecurity lies in resistance, the essence of resistance lies in the trial between the offensive and defensive capabilities of both sides. We should fulfil the cybersecurity responsibility, formulate cybersecurity standards, and clarify protection objects, protection hierarchy, as well as protection measures."

Most importantly, Xi placed cybersecurity at the centre of international great power competition: "the cybersecurity game between large countries is not only a technological game, it is also a game of ideas and a game of discursive power."

The whole-of-system approach is evident in the December 2016 "National Cyber-security Strategy" which relies on the following definition:

> Cyberspace security (hereafter named cybersecurity) concerns the common interest of humankind, concerns global peace and development, and concerns the national security of all countries. Safeguarding our country's cybersecurity is an important measure to move forward the strategic arrangement of comprehensively constructing a moderately prosperous society, comprehensively deepening reform, comprehensively governing the country according to the law, and comprehensively and strictly governing the Party in a coordinated manner, and is an important guarantee to realize … the Chinese Dream of the great rejuvenation of the Chinese nation. (CAC 2016)

Thus, for the Chinese government, three of the essential characteristics of cybersecurity are that it is multidimensional, multi-stakeholder and highly political. This can be further seen in the elaboration by China of the country's cybersecurity priorities based on nine major tasks identified in the strategy (CAC 2016), and summarized in Table 1.1.

The nine major tasks are subordinate to Xi Jinping's "Four Principles" on moving forward reform of the global Internet governance system and the "Five Standpoints" on building a community of common destiny in cyberspace (Xi 2015). Later chapters will address how other stakeholders (corporations, civil society groups, researchers, citizens and investors) address the challenges identified in the strategy. In common parlance in China, the meaning of cybersecurity is more technically oriented than the socio-political approach of the government that it laid out in the December 2016 strategy.

As suggested at the start of this chapter, cyberspace in China is more insecure than in any other major power for a number of technical, social, political, economic, management and social reasons that are discussed through this book. The Chinese government assesses the challenges as "grave" and sees threats to its political stability, economic security, cultural security and social well-being. But it also believes that the opportunities of the digital age outweigh the threats (CAC 2016). The core problem is that China does not yet have a well-developed cyber industrial complex. Some of the nuances and gaps in such official Chinese statements emerge from the analysis in this book. One must never take official Chinese statements at face value.

1.5 National Policy Shift

The public sense of urgency in China around cyber threats reached a new peak in February 2014 when President Xi Jinping announced the country's intention to become a cyber power (Xinhua 2014). He set in train a number of institutional changes to help bring this about. One of the most important was the creation of the CAC during 2014. As indicated in Table 1.1, the current official view in China, as for most wealthy and middle-income countries, is that cybersecurity affects and shapes the full gamut of governmental, economic, social, technological, political and security activities in the country.

Table 1.1 China's cybersecurity priorities

NCSS nine strategic tasks	Functional summary
1. Defend cyberspace sovereignty	*Political security*: protection of the one-party state from cyber-based political subversion or other cyber threats originating inside China or outside it
2. Uphold national security	
3. Protect critical information infrastructure (CII)	*Internal security*: the police and civil emergency responsibility of ensuring resilience of the digital economy and essential services against complex cyber attacks or system failures with cascading national impacts
4. Strengthening online culture	*Cultural security*: protecting digital age information management and exploitation (e-economy, internet culture, countering rumours and fake news, responsible internet behaviour)
… as above	*State administration*: securing a regulated and orderly communications ecosystem in cyberspace
5. Combat cyber terror and crime	*Police and internal security*: protecting corporations, government agencies and individuals from non-political crimes, including malicious cyber attacks or information theft
6. Improve cyber governance	*Cybersecurity social policy*: concepts, enforcement, research, systems development, balancing security and innovation, mobilizing all stakeholders
7. Reinforce the foundations of cybersecurity	*Cybersecurity industry and technical education policy*: S&T foundations, education foundations, industry policy, public awareness, monitoring, risk management strategies and practices, big data policies, cloud computing, internet of things
8. Enhance cyberspace defence capabilities	*Defence policy*: prevent cyber-enabled invasion in peace and war
9. Strengthen international cooperation	*Foreign policy*: diplomacy, norm promotion, capacity building through foreign assistance

The security actions set in train by the government and other actors in China in response both to the "cyber power" ambition and to social and economic opportunities independent of the government remain a rapidly evolving work in progress. This book seeks to address long term trends and so takes a thematic approach in each of its chapters rather than a chronological approach. Nevertheless, at the outset it is useful to have clear view of how important events just before and after February 2014 may prove to be for the evolution of cybersecurity in China.

Information security has been a high priority of the Chinese government since the 1990s (Austin 2014: 34), but as with all countries, it has meant different things to various domestic stakeholders. As a policy objective, it has fluctuated in its ability to hold the attention of the highest level decision-makers. By the end of 2013, there had been a sea change. The leaders were forced to act because of rising concerns about cyber military and espionage activities of the United States, in part revealed by Edward Snowden in June 2013 (Li J 2015; Feng 2014: 4161) but in larger part revealed by increasing effort in the United States to develop its cyber military power (Feng 2014: 4161; Austin 2014: 74–84, 130–45).

The cyber power decision revealed in February 2014 had been made behind closed doors in December 2013. In secret instructions issued that month and subsequently declassified, President Xi asserted ordered the country's specialists to "pay close attention to national security and long-term development, step up planning and formulating a core technology and equipment development strategy and set a clear timetable, vigorously carry forward the spirit of 'two bombs and one satellite' and manned spaceflight, intensify independent innovation" (Chen 2016). He called for "breakthrough points" to be "vigorously supported", and a "focus on the advantages of coordinated research to achieve a breakthrough".

The moves followed unprecedented diplomatic demands on China to curtail its cyber espionage activities for commercial interests (benefiting Chinese corporations). On 12 February 2013, U.S. President Barack Obama, without naming China, alluded to it as an enemy of the United States for seeking to occupy American critical infrastructure through cyber operations (White House 2013). He named financial services, civil aviation and the electricity grid as three primary targets. The remarks came two days after leaks from a U.S. intelligence estimate named China—again—as the most serious menace in the cyber domain (Nakashima 2013). On 11 March 2013, U.S. National Security Adviser, Thomas Donilon, issued three demands on China, which responded the next day saying it was prepared to talk (Asia Society 2013). In response, the Director of National Intelligence, General James Clapper, identified cyber threats to the United States as the number one threat, and talked of a "soft war" against the United States in this domain (Clapper 2013). On 14 March, Obama raised the issue with President Xi Jinping in their first telephone call as heads of state (White House 2013). On 18 March, China's Prime Minister surprisingly called on both China and the United States to stop making "groundless accusations" about cyber attacks against each other (Branigan 2013). On 19 March, U.S. Treasury Secretary Jack Lew discussed the issue when he met Xi in Beijing. One week later, President Obama signed a bill to exclude the purchase of IT products by U.S. government agencies if any part of them is made by a Chinese corporation. For China, the U.S. knowledge of China as the perpetrator was evidence of how poor China had been in protecting the information security of its own espionage efforts.

To drive this process of making China a cyber power, Xi took over the preexisting Central Leading Group (CLG) for Informatization and renamed it by adding the words "cybersecurity" to its title. The office of the CLG (a joint State and CCP body) was also morphed into a new "Cyberspace Administration of China" (CAC) that came into being through the course of 2014. The organization serves both the Party function of supporting the Leading Group (and for that is called the General Office of the CLG), and a governmental function, as CAC, acting as a quasi-ministry under the State Council. The dualistic character of the body is reflected by the fact that at its founding, the Vice Chairs serving under Xi were the Premier (and Chair of the State Council), Li Keqiang, and the CCP's leading official responsible for propaganda at the time, Liu Yunshan. Both were serving as members of the Standing Committee of the Politburo, the seven-man body running the country and the CCP.

The February announcement, backed up with administrative reforms, reflected deep dissatisfaction within the leadership with the pace of innovation in the civil economy. This concern was evidenced by a change in the title of the leading policy group to include cybersecurity as well as informatization. ("Informatization" is not such a common term in English, but as used in China it means application of advanced information and communications technology to all walks of life-political, economic and military. Within the economy, it extends well beyond IT products, such as computers or smart phones, to include the application of information systems in sectors as diverse as health, agriculture, environment and taxation.)

Since then, the government and CCP have been hyperactive on many related fronts: political, legal, economic, military, organizational and diplomatic. By September 2016, with almost three years gone since Xi and his fellow leaders agreed on the new cyber ambition, they were still not satisfied. They announced a detailed new program in line with their five-year plan process, and set a seemingly more robust ambition of becoming a "stronghold" of cyber power (Zhang 2016), in language that the *South China Morning Post* interpreted as the ambition to become a "cyber superpower" (Zuo 2016).

Chinese leaders are not always clear on what the term "cyber power" might mean. At times, they emphasise the foundational aspects: the economic, scientific, technical or military resources they command in cyberspace. At other times, they focus on the dynamic aspects: how well does China perform in its effort to persuade or force other actors to do its bidding in cyberspace or on cyberspace policy issues? The balance between the two (resources versus the exercise of power) in leadership statements reveals a preference for concentrating on power resources to the neglect of power dynamics in cyberspace or on cyberspace issues. Scholarly attention to this subject has for the most part followed that preference-privileging the foundations of power to the relative neglect of the dynamics of China's exercise of cyber power.

The Xi announcement of the cyber power ambition in February 2014 was recognition that China was still lagging in cyber power and cybersecurity and that it was not catching up as quickly as the leaders wanted (Austin 2014: 70). The new ambition has seen many implementing measures since then. For example, in September 2014, Xi told the country it needed a new cyber military strategy. In December 2014, the government introduced new regulations for cybersecurity intended to help promote the rapid growth of China's domestic cybersecurity industry. In May 2015, the country issued a new Military Strategy in which the government declared for the first time in such a document the idea that cyberspace along with outer space have "become new commanding heights in strategic competition among all parties" (State Council 2015). The same month, the National People's Congress (NPC) released a draft bill on National Security (passed in July) that gave a special place to cybersecurity in its provisions for strengthening government control over foreign technologies and related investment in China. Also in July 2015, China released a draft law on network security with sweeping new provisions on control of foreign technologies and data management. The same year, the Chinese armed forces set up a new Strategic Support Force to begin to maximise its advance in practical applications of military cyber capability.

This announcement complemented a series of moves toward institutionalization of new cybersecurity measures in civil sector the previous year that continued into 2016. In March, the government established the Cybersecurity Association of China, comprising 257 members (academic institutes, corporations and individuals) with the mission of promoting self-regulation by industry, entrenching industry standards, deepening of research, and collaborating with international stakeholders (Xinhua 2017a, b). The association is intended to be "a cooperation platform for anyone who is interested in engaging with cyberspace safety", to "play a guiding role in cybersecurity governance, and pursue international exchanges and cooperation".

The move was portrayed by the government as contributing both to national cybersecurity and the country's goal of becoming a cyber power. Its founding chair was Fang Binxing, former President of BUPT credited with being the father of the Great Firewall, who was described in a report in the *South China Morning Post* in 2013 as "one of the country's most hated scientists" (Chen 2013). In November, the national legislature passed the country's first Cybersecurity Law committing the government to "monitor, defend and handle cybersecurity risks and threats originating from within the country or overseas sources, protecting key information infrastructure from attack, intrusion, disturbance and damage". In December, CAC published the country's first National Cyberspace Security Strategy addressing the totality of the problem, from international security aspects to national, corporate and personal interests.

By that time, China had a fairly full suite of new policy guidance and legal foundations in place for cybersecurity. However, when the new network security law came into effect in June 2017, it had become obvious that much work still remained to be done, with the government beginning to lay out regulations and policy statements that would clarify how the new law (essentially just a bare-bones statement of principles) would be implemented.

1.6 Character of China's Political System

The security ambition in cyberspace is heavily influenced by the way in which China's informatization strategy has unfolded. In 2000, China set itself the goal of becoming an advanced information society, where businesses, scientists, and citizens use the most modern information and communications technologies (ICT) to improve performance and enhance social benefit. The CLG for informatization had since 1996 included representatives of key economic and technical ministries, under the direction of a vice premier and with a focus that was mostly economic and scientific/technical. In 2001, it was upgraded and placed under the direction of the premier. In 2002, it became more fully securitized, with heads of security agencies and a peer from the armed forces, becoming active members. In 2014, the leading group was upgraded yet again, being put under the direct control of the CCP general secretary, Xi Jinping. The period since 2002 has seen a massive expansion of the influence of the security apparatus over all aspects of informatization policy in the civil economy.

Chinese leaders regularly advocate a view that the country's political system (one-party authoritarian rule) defines its needs for cybersecurity, especially for what the government sees as internal security. For this reason, "content security" has been a very high priority for cyberspace management in China since the late 1990s. In many respects, the government's high priority to content security (information controls) has almost certainly subordinated or diminished efforts to advance other forms of cybersecurity (across social and technical aspects of control of the eight ingredients as framed by Bell Labs).

China uses a combination of technical and social controls to censor content and control technical activities in cyberspace. The highest levels of the CCP have accepted that technical controls alone are inadequate for achieving their cyber control objectives (Austin 2014: 65) but they invest heavily to maximise controlling effects through technology. The best expression of this position is captured in the term "Great Firewall of China", an analogy with the Great Wall of China. The idea is that the new cyber version of a "great wall" will keep out unwanted intruders in Chinese cyberspace, just as the original great wall was intended to keep out barbarians from Chinese-ruled areas. Chinese officials use this term "great firewall" proudly.

A classic firewall is a "network security device that monitors incoming and outgoing network traffic and decides whether to allow or block specific traffic based on a defined set of security rules" (Cisco 2017). The technology used to block traffic or monitor can be either hardware or software, or a combination of both. The technologies and techniques for applying the controls have changed since they were introduced decades ago and will continue to evolve, possibly quite radically. Chinese researchers openly publish on strengthening the Great Firewall, even though several have been brave enough (while supporting the project in principle) to point out that some of the technical processes involved can have negative impacts on the normal operation of systems.

There is considerable room for Chinese officials and researchers to be careful about the comparison with the Great Wall of antiquity. Waldron (1983) provides convincing evidence of how the 20th century impressions of the purpose, extent and effectiveness of the Great Wall had been manufactured largely by Western historians and chroniclers before they were picked up by Chinese nationalist leader Sun Yat Sen as part of making a new national myth of modern China. The firewall, as this book suggests, may well prove to as porous, as ineffective and as self-defeating as its stone and mortar predecessor. Western cyber analysts need to be as careful as the chroniclers of the original Great Wall and ensure that they do not do the propaganda work of the CCP for it by assuming the Great Firewall to be as effective as its proponents claim. There is a community of researchers, activists and governments outside China who work hard to frustrate the intent and operations of the Great Firewall.

On the social control side, the country now has a vast army of civil servants, uniformed police, co-opted party members and other citizens whom the leaders have mobilized to prevent subversion by cyber means, monitor political expression in cyberspace, and intimidate citizens who may use cyber media to speak out against the political order or simply question any aspect of government policy. This process of close political supervision of the hugely populous country has a dampening effect on the government's efforts to establish regimes for security in cyberspace in other areas of policy that might, at first glance, seem less political: privacy, intellectual property rights, CII, science and technology, and education. But as argued in *Cyber Policy in China* (Austin 2014: 88), the infosphere has its own moral and ethical character that may defeat the "firewall" ambition of the authoritarian government.

Within the state and party structures, there are powerful bureaucratic interests and institutions trying to navigate the new reform paths for national cyber security that have been laid down by the leadership. Many of these are in daily competition to preserve or even expand their bureaucratic positions or increase the flow of budget allocations in their favor. The creation of the CAC in 2014, a joint organization of the CCP and the central government, disturbed the pre-existing power relationships. It was intended to do just that.

As mentioned, CAC was a re-badging of the General Office of the CLG on Informatization. It also incorporated the pre-existing Information Office of the State Council. The CAC had important economic, industrial and technical remits in the field of cybersecurity, but its primary focus was and remains political: that means it spends most effort on content security, including in all of its broader social dimensions. This was revealed through its work program after its establishment but also in the fact that its first head, Lu Wei, simultaneously held the post of Deputy Director of the Propaganda Department of the CCP. Lu was moved aside from his CAC post in 2016, with some sources suggesting it was because he had alienated powerful bureaucratic interests that were competing with CAC. Alasabah (2016) shows how CAC, including its 31 provincial-level offices, have been largely confined to propaganda and content control issues since it was established. This is of course a massive task, given the explosion of social media in China and new technologies of content delivery, especially for video product.

Nevertheless, the CAC (which is also the Office of the CLG on Cybersecurity and Informatization) does have a bureau dedicated to coordinating policy on all aspects of cyber security, which reports to Xi Jinping, the Politburo, and the party's powerful Central Political and Legal Commission (CPLC) and its even more powerful Central Military Commission (CMC). The CAC/CLG office must work with other powerful actors, and the boundaries of their respective responsibilities often shift. Inkster (2015) has a useful overview of Ministry responsibilities in cyber espionage in the period prior to the creation of the CAC/CLG coming into full play.

The most prominent of these other actors that CAC/CLG must deal with are:

- Several units in the Ministry of State Security (MSS) (the MSS reports to the CPLC)
- Bureau of cyber security guard in the Ministry of Public Security (MPS), its 11th bureau (the MPS reports to the CPLC)
- Bureau of science and technology informatization in the MPS (its 22nd bureau)

- Bureau of cyber security management in the Ministry of Industry and Information Technology (MIIT).

Much more is known in the public domain about the MPS compared with the MSS. Western scholars are often keen to delineate sharply the distinct responsibilities of each of these ministries in the field of cybersecurity. (This subject is revisited briefly in Chap. 6.) The publicly available evidence for such distinctions is very thin. Li X (2015: 198) mentions MSS first in his description of "central players", but says its cyber personnel complement is small, dwarfed by that of the MPS' cyber police. But in making this size comparison, Li was not talking about policy weight. We may never know from public sources which of the MSS, MPS or CAC/CLG Central Office has been more powerful in shaping national cybersecurity policy in recent years. We do know that the MSS and MPS both report to the CPLC and its Secretary (chair), who is also a Politburo member, a member of the CCP Secretariat, and one of the most powerful political leaders in China outside the Standing Committee of the Politburo.

The MSS supervises the China National Information Technology Security Evaluation Centre (CNITSEC) established in 2009 to collect information on vulnerabilities of ICT products and information systems. CNITSEC provides this information to the China National Vulnerability Database for Information Security (CNNVD). The latter publishes an annual statistical analysis of vulnerability data and hosts an annual national conference on the subject. A U.S. company, Recorded Future, has reported that CNNVD is a tool of the MSS, used to allow its exploitation of vulnerabilities for espionage or control purposes (Moriuchi and Ladd 2017).

The MPS works closely with the State Encryption Administration (SEA) and the State Secrecy Bureau (SSB), both of which enjoy independent status outside of the MPS. For example, the SSB is essentially the same office as the Central Secrecy Commission which operates under the direct control of the CCP Central Committee.

These non-military agencies also must deal with the relevant parts of the People's Liberation Army (PLA), especially its intelligence arms. The PLA views are reflected to the leadership through a new National Security Commission set up in 2013 by Xi and led by him, through CLGs of the CCP in areas like Taiwan policy, through the CMC (and Ministry of National Defence), and through senior official working groups. After the 19th Party Congress in 2017, the PLA has two representatives in the CPLC, one of whom is the Commander of the People's Armed Police.

There is little public source reporting on how the influence of the MPS has played out in decision-making inside the CAC/CLG but there is enough to make a judgment about key aspects. Several pieces of anecdotal evidence suffice. In 2015, the government released new regulations requiring foreign IT companies supplying Chinese banks to reveal the source codes of the IT products provided. The measure was imposed at the instigation of the MPS to bolster both domestic cybersecurity and localization of the cybersecurity industry. It had an immediate chilling effect on political and economic relations between the United States and China, leading China to suspend the measures in April 2015 after loud protests from the international community, including a personal protest from President Barack Obama. The measure and, indirectly, the negative influence of the MPS, were the subject of a critique by the CEO of Huawei, Eric Xu, who was quoted in April 2015 that China's cybersecurity depends on its openness to foreign technology, not control of it (Shih 2015).

The conservative hand of the MPS has also been seen through the participation of its 11th Bureau in the Internet Society of China (ISOC) since the latter's inception in 2001. ISOC is a multi-stakeholder organization intended to promote maximum economic and social exploitation of the internet while providing a channel for direct political control through the MPS. This ministry has been "first among equals" in the ISOC and, we can presume, always played a constraining role on any cyber policy measures that threatened to dilute control. There is some reason to believe that this "dead hand" led to a certain degree of paralysis in the leading group, resulting in China's slipping off the pace of private sector informatization.

There are a number of technical agencies which support the general remit of cybersecurity in China, and the following information is drawn from their official websites. The most prominent is probably the Computer Emergency Response Team (CERT) set up in 2002, ostensibly as a non-government and non-profit body. It operates as a subordinate body of MIIT. Its formal name is the National Computer Network Emergency Response Technical Team/Coordination Center, and it uses two acronyms: CNCERT (CN for China) or CNCERT/CC. CNCERT has offices in the 31 provinces, autonomous regions and municipalities. The Hong Kong and Macau SARs have their own small central CERTS, which are very different in character from each other. In the former case, GovCERT.HK, set up in 2015, leads on SAR-wide policies and activities as well coordinating emergency information incident response teams in about 80 government departments. It liaises with HKCERT, set up in 2001, which leads work with the private sector and other users. The Macau equivalent, MOCERT, set up in 2009, has only three staff and a very limited remit of technical monitoring, analysis and awareness raising.

The mission of CNCERT/CC is to improve national cybersecurity posture and protect critical information infrastructure. It leads on monitoring, prevention and early warning of threats, and efforts to maximize resilience. In 2016, it received 125,700 incident complaints (compared with peer organization as follows: Japan 16,446, Australia 11,260, and Malaysia 8334) (APCERT 2017: 52–3, 38, 88, 151). This level of reporting in China for incidents is very high compared with Japan (8 times bigger), Australia (roughly double) and Malaysia (four times bigger), though it should be noted that incident complaints may cover different categories of activity in the various countries. It certainly suggests that CNCERT is one of the busiest such organizations in the world.

It was only in 2016 that the government moved to set up a national peak body in this field, the Information Security Association of China. With 257 founding members (corporations, researchers and individuals), its mission is to concentrate resources to support the safety and development of cyberspace (Xinhua 2016). This was followed in 2017 with the establishment of an Information Security Committee mentioned above.

In January 2016, China announced a major reform of its standards committee dedicated exclusively to information security that had been set up in 2002, the National

Information Security Standardization Technical Committee (NISSTC), or Technical Committee 260 (TC260). By the end of 2015, it had published 160 standards in the field drafted "from scratch" (SAC 2016). It has become increasingly active in an effort to keep pace with the new demands of the leadership imposed since February 2014. At the meeting announcing the reforms, the senior Minister attending, Zhi Shuping, levied three new tasks on it (USITO 2016):

- Strengthen strategic planning by drafting standards to complement cybersecurity laws and regulations and by developing a cybersecurity standards roadmap
- Round out the standards system in information security by expedited work on key technologies for critical information infrastructure protection, cybersecurity review, trusted identity, industrial information security, big data security and privacy protection
- Participate actively in international rules and standards making to improve China's influence and increase international adoption of Chinese standards and innovation with Chinese indigenous IP.

Zhi was, at the time, the Minister of the General Administration of Quality Supervision, Inspection and Quarantine, a multifunction agency that supervises technical and security aspects of cross-border activity as well as national level certification and accreditation. It supervises the National Institute of Standardization.

In December 2016, TC260 published a draft set of voluntary standards for public comment addressing cybersecurity, personal information, big data, industrial control systems, and certain other devices. This were intended to clarify some aspects of the new Cybersecurity Law. In June 2107, the committee issued a draft standard on the criteria and procedures for security assessment of data to allow it to be transferred out of China. The composition of the new committee is of special note. It is chaired by CAC Vice Minister Wang Xiujun, with seven vice chairmen representing CAC, MIIT, MPS, CNITSEC, SEA, the China National Certification Administration (CNCA), and the SSB (USITO 2016). The secretary general is the Vice President of the China Electronic Standards Institute (CESI), which hosts the committee offices. It was at the ground-breaking reform meeting in January 2016 that TC260 invited Microsoft to become a member, alongside several other foreign companies (including Intel, IBM and Cisco) as part of an expansion from 48 to 81 members (Microsoft 2016). The committee has seven working groups, addressing specific sub-topics such as encryption and big data.

1.7 Conclusion

China's leaders have strong interest in technical and content control of cyberspace within their borders. They promote development of a cyber industrial complex at home to support their national security policies. On the international stage, they pursue grand strategies for shaping content control in a way that mirrors their domestic interests. The CCP has two over-arching objectives in the public conduct of its cyber-security policies. The first is to convince its own people and the world that it can monitor, shape and dominate cyberspace political content at home as it wants. The second is to develop a national cyber industrial complex that can compete internationally and displace leading foreign actors from most aspects of domestic cybersecurity. Both ambitions are unlikely to be met in the short to medium term. Cybersecurity in China will remain a "joint venture" between, on the one hand, the country's diverse domestic and international stakeholders and, on the other, the powerful and ruthless political leaders running the country. This may be one of the most politically challenging joint ventures the CCP has ever undertaken.

References

Alasabah N (2016) Information control 2.0: the cyberspace administration of China tames the internet. In: Merics China Monitor, 15 Sept 2016. http://www.merics.org/fileadmin/user_upload/downloads/China-Monitor/MERICS_China_Monitor_32_Eng.pdf

APCERT (2017) APCERT annual report 2016. In: Asia Pacific Computer Emergency Response Team. https://www.apcert.org/documents/pdf/APCERT_Annual_Report_2016.pdf

AsiaOne (2017) Nexusguard Places in Top 25 on Cybersecurity 500 List for Eight Quarters in a Row, 21 Feb 2017. http://www.asiaone.com/corporate-news-media-outreach/nexusguard-places-in-top-25-on-cybersecurity-500-list-for-eight

Asia Society (2013) Complete Transcript: Thomas Donilon at Asia Society New York, 11 March 2013. https://asiasociety.org/new-york/complete-transcript-thomas-donilon-asia-society-new-york

Austin G (2014) Cyber policy in China. Polity, Cambridge

Austin G (2015) China's Cyberespionage: The National Security Distinction and U.S. Diplomacy. In: The diplomat, May 2015. http://thediplomat.com/wp-content/uploads/2015/05/thediplomat_2015-05-21_22-14-05.pdf

Bing C (2017) Former U.S. spies say anti-virus software makes for a perfect espionage platform. In: Cyberscoop, 13 Oct 2017. https://www.cyberscoop.com/former-u-s-spies-say-anti-virus-software-makes-perfect-espionage-platform/

Branigan T (2013) China's new premier, Li Keqiang, vows to tackle bureaucracy and corruption. In: The guardian, 17 March 2013. https://www.theguardian.com/world/2013/mar/17/china-premier-li-keqiang-bureaucracy

CAC (2016) China's National Cybersecurity Strategy, issued by the Cyberspace Administration of China. In Chinese. 27 Dec 2016. Original at http://www.cac.gov.cn/2016-12/27/c_1120195926.htm

Chen B (2016) WIN10 misses government procurement. Joint venture will be difficult. In: 21st century business Herald. In Chinese. 16 Aug 2016. http://m.21jingji.com/article/20160816/59622c1aef2877eb1625ec8b526de861.html

Chen S (2013) Great fire wall architect Fang Binxing quits as president of Beijing university. South China Morning Post. http://www.scmp.com/news/china/article/1270213/father-chinas-great-firewall-says-he-will-quit-university-head

China.org (2017) Xi calls for global vision in China's national security work, 18 Feb 2017. http://www.china.org.cn/video/2017-02/18/content_40313020.htm

China Daily (2017a) 网游宵禁 [wǎngyou xiāojìn]: Online game curfew. In English, 07 Feb 2017. http://www.chinadaily.com.cn/opinion/2017-02/07/content_28121149.htm

China Daily (2017b) China to introduce review commission on cybersecurity, 07 Feb 2017. http://www.chinadaily.com.cn/business/tech/2017-02/08/content_28135358.htm

China Digital Times (2017) China reinforces great firewall With new VPN rules, 23 Jan 2017. https: //chinadigitaltimes.net/2017/01/china-reinforces-great-firewall-new-vpn-rules/

Cisco (2017) What is a firewall? In: Cisco website. https://www.Cisco.com/c/en_au/products/ security/firewalls/what-is-a-firewall.html

Clapper J (2013) Remarks as delivered by James R. Clapper, Director of national intelligence, worldwide threat assessment to the house permanent select committee on intelligence, Washington DC, 11 April 2013, https://www.dni.gov/files/documents/Intelligence%20Reports/HPSCI% 20WWTA%20Remarks%20as%20delivered%2011%20April%202013.pdf

Feng D (2014) Preface. China Science Bulletin. 59:4161–4162

Feng S (2017) What is Obama's military legacy. In: Chinamil.com, 17 Jan 2017. The people's liberation army daily (trans), http://www.81.cn/jwywpd/2017-01/16/content_7454421.htm

Hathaway M, Klimburg A (2012) Preliminary considerations on national cybersecurity. In: Klimburg A (ed) National cybersecurity manual. NATO CCDCOE, Tallin, pp. 1–43

Inkster N (2015) The Chinese intelligence agencies. In: Lindsay J, Cheung TM, Deveron DS (eds) China and cybersecurity: espionage, strategy, and politics in the digital revolution. Oxford University Press, pp. 29–50

Li J (2015) Network information security challenges and relevant strategic thinking as highlighted by "PRISM". In: Huang Z, Sun X, Luo J, Wang J (eds) Cloud computing and security. Lecture notes in computer science, vol 9483.Springer, Heidelberg, pp. 147–156

Li X (2015) Regulation of cyber space: An analysis of Chinese law on cyber crime. Int J Cyber Criminol 9(2): 185–204 (July–December)

Marro N (2016) The five levels of cybersecurity in China. China Business Review. 5 December 2016. http://www.chinabusinessreview.com/the-5-levels-of-information-security-in-china/

Mi Y (2017) Public's leaked personal information seized in online safety campaign. In: China daily, 5 Jan 2017, http://www.chinadaily.com.cn/regional/2017-01/05/content_27870730.htm

Microsoft (2016) China invites Microsoft to join technical committee 260 (TC260) to draft cybersecurity rules. 26 Aug 2016, https://mspoweruser.com/china-invites-microsoft-to-join-technical-committee-260-tc260-to-draft-cybersecurity-rules/

Moriuchi P, Ladd B (2017) China's ministry of state security likely influences national network vulnerability publications. In: Recorded future blog, 16 Nov 2017, https://www.recordedfuture. com/chinese-mss-vulnerability-influence/

Nakashima E (2013) U.S. said to be target of massive cyber-espionage campaign. In: Washington post, 10 Feb 2013. https://www.washingtonpost.com/world/national-security/us-said-to-be-target-of-massive-cyber-espionage-campaign/2013/02/10/7b4687d8-6fc1-11e2-aa58-243de81040ba_story.html?utm_term=.11552f05b1cc

SAC (2016) Standards administration of China national information security standardization technical committee general assembly held in Beijing, 15 Jan 2016, http://www.sac.gov.cn/xw/bzhxw/ 201601/t20160115_200544.htm

Shih G (2015) Huawei CEO questions China's cybersecurity policies. In: Christian Science Monitor, 22 April 2015, https://www.csmonitor.com/Technology/2015/0422/Huawei-CEO-questions-China-s-cybersecurity-policies

State Council (1994) Regulations for safety protection of computer information systems. Decree No. 147, 18 Feb 1994. Available at AsianLII, http://www.asianlii.org/cn/legis/cen/laws/rfspocis719/

State Council (2015) Full text: China's military strategy, China daily, 26 May 2015, http://www. chinadaily.com.cn/china/2015-05/26/content_20820628.htm

USITO (2016) TC260 new committee opened to foreign participation. http://www.usito.org/news/ tc260-new-committee-opened-foreign-participation

Waldron A (1983) The problem of the great wall of China. Harv J Asiatic Stud 43(2):643–663

White House (2013) Readout of the president's phone call with Chinese president Xi Jinping, 14 March 2013. https://obamawhitehouse.archives.gov/the-press-office/2013/03/14/ readout-president-s-phone-call-chinese-president-xi-jinping

Xi J (2015) Remarks by H.E., Xi Jinping president of the people's republic of China at the opening ceremony of the second world internet conference, Wuzhen, 16 Dec 2015. http://www.fmprc.gov.cn/mfa_eng/wjdt_665385/zyjh_665391/t1327570.shtml

Xi J (2016) Speech at the national meeting on cyber security and informatization. In Chinese. 25 Apr 2016. http://www.xinhuanet.com/politics/2016-04/25/c_1118731175.htm

Xinhua (2014) Xi Jinping leads internet security group, 27 Feb 2014. http://news.xinhuanet.com/english/china/2014-02/27/c_133148273.htm

Xinhua (2017a) Thousands of illegal apps taken offline in south China, 24 Jan 2017. http://news.xinhuanet.com/english/2017-01/23/c_136007345.htm

Xinhua (2017b) Procuratorates approve arrest of 19,000 telecom fraud suspects, 14 Jan 2017. http://news.xinhuanet.com/english/2017-01/14/c_135982423.htm

Xinhua (2016) China's first national NPO in cybersecurity founded, 25 March 2016. http://news.xinhuanet.com/english/2016-03/25/c_135223674.htm

Zhang S (2016) China sets goals of informatization, CRI news. 28 July 2016. http://english.cri.cn/12394/2016/07/28/3821s935816.htm

Zuo M (2016) China aims to become internet superpower by 2050, South China morning post, 28 July 2016. http://www.scmp.com/news/china/policies-politics/article/1995936/china-aims-become-internet-cyberpower-2020

Chapter 2
Education in Cyber Security

Abstract This chapter sketches the terrain of China's university-level education in the cybersecurity sector. The chapter also looks briefly at the impact of an internationally mobile work force on China's talent development.

2.1 Formal Education as the Key

There is considerable room in the cybersecurity sector for inventors, entrepreneurs and inspired (white hat) hackers who do not have formal education qualifications in some closely related field. The larger part of the sector however depends on high quality basic research by scientists, including social scientists. Educational institutions are the primary sources of the most skilled cybersecurity workers, teachers and trainers. As this chapter shows, the official position of the Chinese government, for better or for worse, is that Chinese universities are the main drivers of talent development of Chinese people in all realms of cybersecurity because they provide the teachers and the trainers in both university settings and in other, non-university teaching and training environments. At the same time, the government and Chinese corporations pay significant attention to two-year colleges, the need for continuing education, refresher courses, on the job training and lifelong learning. In addition to China's 150 or so more capable universities, there are 850 additional higher education institutions in the country, though only a small selection address cybersecurity education or training.

Cybersecurity technologies and the related workers are globally mobile. Yet China (like most governments) tries to carve out sovereign capability in areas of cybersecurity or activity which it believes is essential to national security. Most governments look to reserve knowledge of certain dimensions of a country's cybersecurity to a select group of their own citizens while other governments seek to steal that knowledge, most often by cyber espionage or simply direct recruitment of the personnel involved.

© The Author(s) 2018
G. Austin, *Cybersecurity in China*, SpringerBriefs in Cybersecurity,
https://doi.org/10.1007/978-3-319-68436-9_2

For all of these reasons, to arrive at an understanding of China's scientific and educational base for cybersecurity is a highly complex, multi-faceted challenge. This short chapter can only scratch the surface of the situation. The economics, politics and cultural dimensions of the domestic education and research institutions (predominantly a "nationals-only" enterprise) are quite distinct from the blended international and domestic character of the work force in China. Even in looking at the work force, there are two quite distinct approaches.

Taken together, these various perspectives on the political economy of knowledge, skills and abilities in the cybersecurity of Chinese government agencies, corporations and individuals represent both a massive research target and a largely unstudied one in the public domain inside the country or internationally.

2.2   Nature of the Problem

It would be great if the challenges of cybersecurity education, digital literacy promotion and workforce development could be easily bounded and managed by a single government agency or peak professional body in the sector (Austin 2017). The challenges cannot be so bounded. Research on security in the information age and the teaching of its various aspects are as unbounded a set of problems, and as multidimensional, as we can find elsewhere in social policy short of problems like ending crime. The United States, which is the most wealthy country in the world and the most capable cyber power, has failed to meet the cybersecurity education needs it has set itself, with President Trump announcing new urgent measures early in his Administration (White House 2017). This followed moves by President Obama to declare a national emergency in cyber space for two years running in 2015 and 2016. The Trump Executive Order introduced a new concept of a cyber workforce arms race, calling for a review of "the workforce development efforts of potential foreign cyber peers in order to help identify foreign workforce development practices likely to affect long-term United States cybersecurity competitiveness".

Authoritarian governments such as China, have been no more capable than the wealthiest democracy, and arguably less so, in mobilizing adequate cybersecurity education strategies to meet their work force needs. In 2011, a Chinese researcher, Luo (2011: 294), observed that time had not been on China's side. Noting that relevant university courses had only recently been set up, Luo observed that the country also suffered from a deficit in trained educators in the field, especially people who might be considered pace-setters.

This is borne out by the trends in number of PhD graduates in cyber security in China which has shown an unusual downturn after 2011, according to data based on how graduating students described their work in submitting it to the China Knowledge Resource Integrated Database (CNKI). Figure 2.1 shows completions recorded in the CNKI database for the period from 1999 to September 2016 where the successful PhD scholar identified information security or cybersecurity as a subject of the dissertation (Lü 2017: 3–4). The data may be incomplete, especially in earlier years, and it

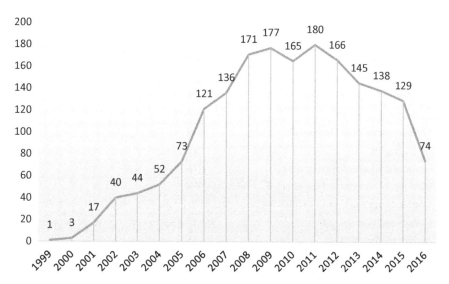

Fig. 2.1 Recorded PhD completions in information security by year 1999–2016 (part year)

contains few completions from Beijing University or Tsinghua University since these universities have a policy of ignoring this database. The number of cybersecurity PhD completions from these universities in any year is not large. The peak year appears for the whole country to have been 2011, and the peak five-year period appears to have been 2008–12. The average completion rate in Chinese universities for the 12 years since 2005 inclusive represents not much more than one new PhD graduate per year per university now offering undergraduate majors in cybersecurity. This rate of PhD graduation provides a very weak foundation for large scale expansion of education in this sector, even if one assumed the very unlikely case that all PhD graduates in the field go into university teaching posts and none go into other forms of employment.

One explanation for the decline in Chinese home-grown PhDs in the field is that doctoral completions by Chinese citizens in the United States in all engineering and science fields increased by 50% between 2005 and 2015 (NSF 2017: Table 26). While specific statistics are not available in the NSF data, it is possible to imagine that China has chosen to invest more in doctoral education of its nationals at U.S. and other foreign universities than in its own universities since 2011. On the other hand, perhaps the labour market is drawing people to study in the United States, since the stay rate of China-origin PhD graduates in the United States in all disciplines (at least expressed as an intention) was 86% in the period 2005–2015 (NSF 2017).

In terms of discipline specialization, China's PhD completions in information security have not reflected a commitment to the multi-disciplinary vision of the 2016 National Cybersecurity Strategy referred to in the previous chapter. According to the available CNKI data, only 9% were undertaken in the social sciences, and among the rest, there was an overwhelming concentration in engineering (1251 out of 1854–or two thirds). That heavy imbalance may reflect the practice in Chinese universities

Table 2.1 Recorded PhD completions in sub-specializations by year 1999–2016 (part year)

Sub-specialty	Number	% share
Computer applications technology	394	21
Communication and information systems	234	13
Computer software and theory	133	7
Computer architecture	127	7
Computer science and technology	109	6
Cryptography	91	5
Signal and information processing	73	4
Management science and engineering	69	4
Control theory and control engineering	59	3
Circuits and systems	37	2
Applied mathematics	35	2
Information and communication engineering	29	2
Power system and automation	22	1
Traffic information engineering and control	19	1
Optics	17	1
Microelectronics and solid state electronics	16	1
Mechatronic engineering	13	1
Optical engineering	12	1
Physical electronics	12	1
Mechanical design and theory	12	1

of awarding degrees that study information security under the discipline name of engineering even if the subject was not classic electrical engineering. Table 2.1 shows the breakdown of the top twenty identified sub-specializations (Lü 2017: 9).

These considerations in respect of tertiary level programs sit alongside a myriad of other fundamental curriculum issues that any country faces in reorienting its educational and vocational training system to adjust to the threats of the cyber age. Outside China, researchers have noted fundamental weaknesses even in the very conceptualization of cybersecurity education (Austin 2017). Schneider (2013) suggested that change is needed on all fronts in how universities develop curricula in this field and deliver them. Conklin et al. (2014: 2008) have warned against any vision of cybersecurity education that is monochromatic: "The development of a single foundational curriculum that can meet all major requirements is not a possibility for a field as diverse as information security. Information security is a field that has both breadth and complexity."

Austin (2017) identifies a list of dilemmas or serious education and work force challenges or dilemmas that exist in most countries, and these are highly relevant for China:

1. Weak evidence-base for policy (who is researching cybersecurity education?)
2. Education policy choice points beyond the multiple curricula (resources, quotas, incentives)
3. Need for a system to measure maturity in education systems for cybersecurity (baseline criteria)
4. The role of immigration and off-shore labour
5. Political economy: cost transfer of training from the private sector to the public sector
6. Online education and training: international and domestic
7. Disruptive technologies; impact on curricula
8. Formal knowledge and education versus self-taught informal knowledge
9. Critical thinking, adaptability and personal resilience as the core abilities in cybersecurity.

In respect of the first point above, it is worth noting that in the two decades to 2016, there has been only one recorded PhD completion in China in the field of basic education and secondary vocational education for cybersecurity.

The remainder of this chapter will provide some further insight, first by looking at the views of the Chinese government. This is followed by a review of university approaches to cybersecurity research. The chapter then offers some comment on the activities of leading Chinese universities, based on a separate study (Austin and Zu 2018). The chapter then turns briefly to the place of overseas universities and international professional associations, before taking a closer look at several workforce development issues.

By way of introduction, it needs to be noted that China's education and knowledge economy faces severe structural constraints (Austin 2014: 20–48, 89–128). First, this is the biggest education system in the world. Second, the country (government and people alike) sets itself high, possibly unrealistic ambitions in respect of carving out Chinese leadership and "Chinese characteristics" in the globalized knowledge economy of cybersecurity. Only 26 universities are approved to offer MoE scholarships for foreign students, and this represents a severe constraint. Third, the country has a highly centralized education management system for universities which allows for targeted strategic campaigns in certain privileged sectors (OECD 2016: 12) but it also stifles creativity in others not lucky enough to be selected. Fourth, the university sector in China was all but disbanded for around a decade because of the Cultural Revolution (1966–76), a violent political upheaval which resulted in millions of deaths, the persecution of intellectuals as "stinking weeds", and mandatory rural labour for most of them as a substitute for university work. This period came to an end just over 40 years ago, but was "revisited" with chilling political effect just a decade later. Country-wide student demonstrations between 1986 and 1989 culminated in the massacre of students in TianAnMen Square on 4 June 1989 and the launch of an ideological rectification inside universities. Fifth, the CCP still imposes a second layer of bureaucracy on an already ponderous and numerically enormous system through its insistence on having a system of political commissars (CCP committees) inside scientific and educational institutions (see Article 39 of the higher Education Law of China). In 2014, the CCP began a campaign against foreign influences and books in universities, with President Xi calling in 2016 for them to become bastions of the CCP. Academic promotions in Chinese universities depend on CCP activism. Sixth, China has a weakly developed domestic cybersecurity industry which impacts the push and pull factors for skill development in the labour market, though massive increases in demand have arisen in recent years. Seventh, China does not yet have a highly developed national innovation system based on the "triple helix" of government-universities-private sector collaboration which has been seen as the root of knowledge power and high tech industry development elsewhere, especially in the United States (Austin 2014: 89–128).

2.3 Government Policy

China's Ministry of Education approved the first undergraduate degree in information security in 2001 (Hui and Tan 2016: 13). This was thirty years after the first (quite basic) university courses in "computer protection" began to appear in the United States (Hartson 1983) and the publication of the first issue of the first major academic journal in the field, *Computers and Security*. It was not until 1993 that China even set itself the goal of education reform to build a small number of world class universities out of a pool of around 100 key national institutions (Luo 2013: 168). Nevertheless, China's PhD programs in all disciplines have produced about the same number of graduates each year in the recent past as U.S. universities: 54,070 in 2014 for the United States (NSF 2015: 2) and 52,654 for China in 2015 (MOE 2016a).

After 2001, and a string of other important policy turning points early in the decade, China had been recording steady improvements in cybersecurity education outcomes but they did match either the character of the national problems or the leaders' great power ambitions. In 2005, only 47 tertiary institutions in the country had a major in information technology, and even fewer had majors in cybersecurity. The Xi announcement in February 2014 of the country's intent to become a cyber power and the institutional reorganizations in security affairs around that time, including in military policy, presaged a quickening of the pace of cybersecurity education. The 13th Five Year Plan (2016–2020) includes a science and technology program to 2030 which will concentrate on six areas, including cybersecurity and quantum communications for its security applications (Yu 2017: 1759).

The terrain for education reform in favour of cybersecurity would prove to be problematic. As one recent study observed: universities retain conservative attitudes when it comes to information security as a discipline (Wang et al. 2017: 84). These authors report low confidence in the commercial sector about the quality of university education in this field. Moreover, in addition to the systemic constraints mentioned above, the drive to create world class universities in China must win the acceptance of the affected faculty and students if there is to be any effective implementation (Kim et al. 2017: 6). Such a critique implies a process of protracted and episodic reform in China as it seeks to expand cybersecurity education.

The basic direction of current reform efforts in this field was revealed on 29 November 2014 by Feng Huaming, the Secretary General of the Information Security Education Steering Committee and Vice President of Beijing Institute of Electronic Science and Technology (Xinhua 2014). The tasks set were to: "have a number of high-quality information security management and technical personnel, to strengthen the information security disciplines, professional construction, and speed up the training of information security personnel". Of these four tasks, the author believes that the biggest challenge for the country was probably going to be expanding the pool of highly qualified people at all levels of information security management. The other tasks would not be easy.

Feng correctly situated China's needs within the context of the "optimization of global resources and the development of innovative models" characteristic of the ICT sector as whole against the background of intensifying competition and increasingly serious threats and vulnerabilities. He echoed the centrality of network and information security to national security. He cited a 2003 assessment (Document 27) that "China's information security work, network and information system protection level was not high", that the country lacked information security management and technical personnel, and that national information security awareness was weak. He said that this document identified the same tasks that the country was now setting itself. This was followed, he said, by a Ministry of Education (MoE) plea (Document 7) in 2007 to colleges and universities to bolster research and education in the field. The same year, the Ministry of Education set up an Information Security Professional Education Steering Committee to lead policy for higher education institutions in order to promote development of the field. In 2012, the State Council (Document 23) called for more vigorous support for the relevant disciplines. The February 2014 meeting at which Xi made the "cyber power" announcement included similar, more forcefully expressed invocations: "strengthen information security personnel construction, create world-class scientists, network technology leaders, excellent engineers, high level innovation teams".

Feng referred to specific policies of the United States and the United Kingdom to build their workforce as a "revelation" to him. In the U.S. case, he said, as early as 2005, the U.S. National Security Agency commissioned more than 50 educational institutions (including Carnegie Mellon and Purdue University) to set up academic centres to build the tertiary sector's personnel training capacity in information security. He also noted how Microsoft, Cisco, Boeing, GM, and other enterprises had built their own systems of information security training. He said the Snowden incident in 2013 "fully demonstrated that the cyberspace struggle was complicated and intense" and that China needed a national innovation system for information security that could help mount collective defences.

Feng laid the responsibility for building this system at the feet of colleges and universities, saying that the country needed both national security strategy guidance and the policy support for the discipline". He reported that the country's efforts at workforce construction had not been systematic or concerted. He noted that the system of discipline management in Chinese institutions had resulted in low allocation of resources. Noting that military institutions had a long history in the field, especially in cryptography, he said that it was only in recent years that other institutions came gradually to recognise the importance of network and information security, and therefore to the importance of related construction of the discipline. But, he said, they had failed to build the discipline adequately. He recommended that as soon as possible China should declare network and information security related disciplines be established as a primary (first level) discipline.

The attention paid by Feng to China's active participation in international competition for information security personnel was admirable, if possibly atypical in the emerging techno-nationalist environment. He said that "governments can use tariffs or barriers to protect their own industries and to control the flow of factors of production, but they cannot control the flow of talent". He observed that foreign developed countries are actively seeking out China's talent and hiring them. He reiterated that China "certainly cannot decide to close the country and cannot limit the flow of scientific and technological personnel". He called on the Central Organization Department of the Central Committee of the CCP to attract overseas high-end talent to China.

As of 2014, 80 universities and colleges had set up majors in information security (military universities not included), of which 17 had set up specialization in information security countermeasures, and 10 had established departments specialising in privacy (Wang et al. 2017: 85). By April 2016, the numbers had grown to 122, 18 and 12 respectively. These programs can train 10,000 graduates about per year (Zhang et al. 2016: 5). The figure given by Feng Huaming for 2016 mentioned 143 internet security majors spread over 126 universities (Xinhua 2017). On this author's estimate, there are probably around 20 military and security related universities with information security majors, though the majority of these are not first tier universities, but rather are provincial police universities.

As indicated by Wang et al. (2017: 86–7), the study of information security in Chinese universities suffered from its failure to be recognized as a "first level discipline". It was covered within 14 existing first level disciplines, almost all of which were technical and had a heavy emphasis on cryptography. (In 1997, China set up a system based around 12 disciplines, 88 first-level disciplines and 382 second-level disciplines.) According to the study by Zhang et al. (2016: 88), there were several shortcomings in information security courses when it was not a first-level discipline, such as subordination to the first-level discipline (less ability to specialise in information security), lack of practical content, low investment in human resources for teachers, and inability of graduates to get jobs in the field.

It was not until June 2015 that the MoE and the State Council Academic Degrees Committee decided to establish cyberspace security as a first level discipline (Hui and Tan 2016:13). Graduates majoring in this discipline would be awarded an engineering degree. In October 2015, the two agencies also approved a first level discipline designation for doctoral studies in cyberspace security. Both measures were announced in January 2016, authorising some 27 universities, including three military universities (National Defense Science and Technology University, PLA Information Engineering University, and the PLA Polytechnic University) to implement both steps (MoE 2016b). The decree enjoined the universities to "further integrate resources, strengthen discipline construction, and do a good job of cyberspace security personnel training work".

One implication of being a first level discipline is that the MoE maintains a teaching guidance committee for each specialization with that designation. Without being a first level discipline, a field can still attract significant investment, as had been the case with modelling and simulation–for which MoE approved 100 new experimental centres in 2016 (Zhang et al. 2017: 169). This is a high priority education field for cybersecurity. Computer science (including information security) is one of the top five fields for simulation in China, and is heavily funded by the PLA, an aspect which defies easy open source documentation of developments in education policy. It was also in 2016 that a new academic journal was launched, the *Journal of Cybersecurity*, sponsored by the Institute of Information Engineering Research in the Chinese Academy of Sciences and the China Science and Technology Publishing Media Company. In 2016, President Xi Jinping led a work conference on cybersecurity at which education was a primary theme.

As mentioned above, in March 2017, President Donald Trump ordered a review of U.S. cybersecurity education and workforce needs. By August 2017, China followed suit and announced a new education strategy in the field (CAC and MoE 2017). Its strategic goals were to build world class cybersecurity research and education centres by 2027 as demonstration projects for the rest of the country and to give the lead to local governments and corporations on how they must participate in the development process. The vehicle would be a funding scheme jointly managed by the CAC and MoE which would be subject to application and annual review of performance by the two sponsoring agencies. This is similar to the UK scheme for centers of excellence in universities.

The ambitions for China were modest and well-phased: "After about ten years of trying, 4–6 internationally and domestically influential and well-known cybersecurity institutions would be established". While China may be considered better off than most countries in the world, it is definitely not in the top 20 in cyberspace security and it currently has no "internationally and domestically influential and well-known cybersecurity institutions" for education and research. Missing from the official announcement was any sense that the subject is highly diverse, and that each centre might reasonably aspire only to becoming a power house in several discrete sub-fields, some in technology, some in cryptography and some in social, legal and management aspects. Moreover, the balance in Chinese policy as of 2017, including education policy, was still very heavily in favour of the technical dimensions of cybersecurity rather than a deep appreciation of it as a socio-technical phenomenon. Some Chinese professors point out that the government's concept of world class universities is one that does not take account of Chinese realities and that it is too dominated by Western managerialist approaches emphasising research outputs (Kim et al. 2017).

2.4 Information Security Research

Chinese researchers have claimed world-leading breakthroughs in certain areas of research relating to information security. This is the case with quantum computing, which researchers believe holds out the opportunity for "highly secure computing" as an alternative to existing systems that have high inherent insecurity. For example, in July 2017, as captured in headlines from the *MIT Technology Review*, China teleported the first "object" from earth to orbit (Tech Review 2017), and according to *Science Magazine*, China's quantum satellite achieves 'spooky action' at record distance (Popkin 2017). The later referred to the breakthrough as a "stepping stone to ultrasecure communication networks and, eventually, a space-based quantum internet". The original scientific paper was published in arXiv (Ren et al. 2017).

However, in general, China's premier universities still lag behind their leading Western peers in terms of research quality and international impact (Shuang et al. 2016: 315). This is the case for the discipline of computer science and engineering, including information security (Olijnyk 2014; ATIP 2015). Olijnyk (2014: 164) found, relying on simple keyword search, that while Chinese authors produce more works in the field of information security than their U.S. counterparts, they significantly underperform in terms of impact on the field (as indicated by very low citation rates for their work). Olijnyk's data shows, though he did not mention it, that apart from the work done in China, the near totality of research in his data sets is conducted in countries that are allied with the United States. The United States sits at the epicenter of a globalized and dominating epistemic community for information security that is oriented towards shared values, many of which China opposes.

Shuang et al. (2016: 318) recognize that "the research level of some Chinese universities can be classified as world's first class", but they see particular problems of gaps in the "development balance" of major disciplines. The authors' recommendations (319–20) include:

- "Zero in" on the frontier of the discipline and improve the quality of academic papers
- Capitalise on international academic exchanges
- Accelerate the development of new disciplines and interdisciplinary study
- leading universities need to invest heavily in subject areas in which they can excel rather than trying to be good at everything.

The cross-disciplinary study of information security in Chinese universities is only in its infancy—a "new discipline" (Zhang et al. 2015: 2). This proposition can be supported by a comparison of curriculum content between Chinese and U.S. universities in the field (Chen et al. 2013). The discussion below of specific Chinese universities bears out the continuing relevance of that 2013 finding.

According to Zhang et al. (2015: 3), traditional information security looked mainly at data security, whereas the new discipline in China of "cyberspace security" now had to include equipment safety (hardware security), content security and behavioural security. Their analysis of the research scene in China and globally (with some 287 references) and their view of future research tasks pays almost no attention to management and behavioural aspects. In that area, they called for new research into information security management platforms for large networks. The paper calls out five challenges in future applied research (mobile terminals, network devices, software-defined networks, cyber physical systems, and the 5G network). In addition, it gives extensive attention to trusted computing and cloud security. There is a four-page treatment of content security ("adverse information on the internet") (28–32). The authors note specific shortcomings in the current practices and research in this field, as they did earlier for the indigenously developed Kylin operating system introduced quickly in an effort to replace Microsoft Windows. The paper reports the contribution of scientific research to the development of three information security standards in China in 2013: motherboard functions and interface of trusted platforms, trusted connection architectures, and functionality and interface specification of cryptographic support platforms for trusted computing. In the early pages of the article, the authors pay considerable attention to cryptography, an area that may be one of the strongest sub-fields in China in information security. The paper is particularly valuable for demonstrating the appreciation in China of the sheer vastness and complexity of the multi-disciplinary foundations of information security research.

While in general terms, it might be possible to say that China's research capability in information security lags behind that of the United States, the Chinese research community is vigorously engaged with its peers globally. Leading institutions have hosted international conferences, beginning for example in 2005 with the first Conference on Information Security and Cryptography organized by the State Key Laboratory of Information Security (SKLOIS) of the Chinese Academy of Sciences (CAS). The proceedings were published in English by Springer, based on 196 submissions from 21 countries that were reviewed by the organising committee to produce a list of 33 accepted papers and another 32 short presentations (Feng et al. 2005). China has hosted a large number of the annual conferences on Information Security Practices and Experience (ISPEC series) beginning with the second conference in 2006, with proceedings also published by Springer. In assessing the research performance of its scholars, leading Chinese universities now place a premium on publications in leading international journals.

2.5 Selected Research and Education Institutions

In 2017, the Chinese Universities Alumni Association (CUAA) released a ranking of the best universities in China for information security, assessing four at equal first place with a 6-star rating, and another fourteen at 5-star (with 9-star being world leading), as shown in Table 2.2 (CUAA 2017). Like any university, a ranking of the discipline of this kind does not speak to the upper end talents of its best achievers, but it does speak to the likely volume of such high end talents. What is notable here is that only four universities are assessed at the six star level in this discipline (a high level by world standards, best in China).

Table 2.2 CUAA rankings of best universities in the undergraduate level information security discipline

Rank	Name	Rank	Name	Rank	Name
1	Wuhan U.[a]	2	Nanjing U. of Posts & Telecommunications	3	Xidian (Xian)[a]
1	China U. of Electronic Science & Technology[a]	2	Beijing U. of Posts and Telecommunications[a]	3	China U. of Science and Technology
1	Shandong U.	3	Beijing U. of Aeronautics and Astronautics	3	Northeast U.
1	Shanghai Jiaotong U.[a]	3	Harbin Institute of Technology[a]	3	Central South U.
2	Hangzhou U. of Electronic Science &Technology	3	Sichuan U.[a]	3	Fudan U.
2	China Criminal Police College	3	Huazhong U. of Science and Technology		

Of these, a short research briefing prepared in conjunction with preparation of this book (Austin and Zu 2018) gives an overview of seven, which are marked with [a] in Table 2.2. Beyond the above list, the briefing looks at Tsinghua University and the State Key Laboratory of Information Security in the Academy of Sciences (SKLOIS). It is remarkable that neither Tsinghua nor Beijing University, the country's two premier education and research universities, appear in the rankings at all. The centre of gravity in Chinese studies of information security broadly defined is not in Beijing, and nor is it in Shanghai, though these cities are prominent. Summary highlights as assessed by this author are as follows:

- Few universities have adjusted to the holistic view of cybersecurity as a socio-technical phenomenon, with almost all being heavily oriented toward computer science, mathematics and engineering. The main exception is Xidian University in Xian which has close links to the PLA and was probably ordered and resourced to make the switch. Other pacesetter universities include Wuhan and the University of Electronic Science and Technology in Chengdu
- The numbers of staff teaching cybersecurity do not appear to have grown dramatically, and in an institutional sense seem to remain under the shadow of the larger computer science, engineering or maths departments
- The numbers of graduates in cybersecurity remain low relative to the demand, and there has been only modest growth in intake of new undergraduates since 2014 in some institutions. Several leading cybersecurity universities have seen declining numbers of new entrants in that time frame
- Few universities have advanced simulation capability for cyber security teaching and few have strong links with private industry in this sector.

2.6 Overseas Universities and Research Institutes

China's skills base in cybersecurity is highly dependent on both Chinese and foreign specialists who may have been educated in universities overseas. So, the Chinese government and other stakeholders must (as most of them do) have a view on the balance between Chinese information security specialists educated abroad or at home; and separately, a view on the balance between Chinese citizens (regardless of where they were educated) and foreign specialists working for Chinese entities, either inside China or outside it. In addition, China relies to some degree on cybersecurity services provided by specialists who never set foot in China but who work remotely with domestic partners. As one of the wealthiest countries in the world China can afford to buy offshore cybersecurity services, including education services, on a scale that few countries can match.

It is widely known that university departments outside China specialising in the many theoretical and practical aspects of information security, especially in the technical aspects of the field, enjoy significant representation from Chinese citizens. For example, the quantum computing department at MIT has several leading Chinese scholars. However, it is in the nature of Western universities not to keep track of the nationality of their researchers, though universities and individual scholars in countries like the United States are required to observe certain security practices to prevent the flow of protected information about a small range of sensitive military or dual use technologies.

Since content control is a high priority for China's cybersecurity and since most Western universities do not teach it, Chinese graduate students with this interest (often mid-career officials in relevant security agencies) can choose (or be sent) to study abroad in countries like Singapore and the United Arab Emirates, the governments of which actively restrict internet content with undesirable political implications. One of the better known centres in this field is the Lab for Media Search (LMS), a multimedia research group in the school of computing at the National University of Singapore (NUS).

2.7 International Professional Associations

One of the most important globalising forces in China's cybersecurity workforce and worldview may be the international professional associations, such as the Institute of Electrical and Electronics Engineers (IEEE), ISACA and ISC^2. Of these, IEEE has been and remains the most influential in many areas of technical and social development of cybersecurity. China will know it has arrived at global leadership in information security education when its professionals can set up genuine NGOs like ISACA and ISC2 that have such global impact.

According to data on various IEEE websites, a Beijing section of IEEE was set up in China in 1984 as the country launched its big push into an electronic future with the appointment of Jiang Zemin as Electronics Minister. In 2007, IEEE set up its China office, and by the end of 2012 could count 10,000 members in China. The office is very active in organising conferences for its members, including in areas of policy and management. The meetings in China, usually around 150 per year over the last decade, have attracted tens of thousands of delegates. By December 2013, another 2000 members were added to the IEEE in China, and the organization had seven sections (Beijing, Shanghai, Nanjing, Wuhan, Xian, Harbin, Chengdu), reflecting the electronic industrial hubs in China. While cybersecurity is only one of many professional interests for IEEE members in China, the organization operates at the forefront of policy and ethics in this field in China.

The Hong Kong chapter of ISACA (previously the Information Systems Audit and Control Association), entered China in 2009 to create a joint China Hong Kong chapter (ComputerworldHK 2009). At the time, it claimed 1700 mainland members out of global membership of 86,000, about 2.5%. The organization is dedicated to professional leadership and certification in information systems audit, security and governance. By 2017, ISACA had almost 500,000 members in 188 countries, organized in 215 chapters, and an office in China, as well as its headquarters in the United States. Its new Board appointed in mid-2017 includes members from Singapore, Japan and India but none from China.

The world's leading certification organization for information security professionals, (ISC)^2, set up its China office in 2013 to further develop its reach into that country, according to its website. Apart from supervising the certifications, it runs programs, including in China, such as an Authorized Instructor Program, an Official Training Provider Program, and a program for Youth Internet Protection Safe. (ISC)^2, set up in 1989, claims more than 123,000 certified members in over 160 countries.

2.8 Workforce

In cybersecurity, the work force is highly diverse: from engineers and scientists to keyboard operators, accounting personnel and management roles; from product developers and sales people to intelligence analysts, forensic specialists, cyber attack planners; from CEOs and directors of companies to government policy managers and officials leading standards development. There are many levels of capability, ranging in broad terms from basic through intermediate, advanced and super expert. Knowledge sets for cybersecurity can be a single discipline (such as electrical engineering, management or law and policing) or be multi-disciplinary, since cybersecurity is a socio-technical phenomenon.

China may have the largest cybersecurity skills deficit of any country in the world. This shortfall is expected to be around 1.4 million by 2020, according to Feng Huaming, a Vice president of the Beijing Institute of Electronic Science and Technology Institute (Xinhua 2017). It is highly unlikely that the Chinese education system, training networks and labour market will begin to fill this gap for one or two decades at least, if the country relies largely on home-grown talent. In quantitative terms, the estimated size of the new graduate population each year in the field is about 10,000. While a tertiary degree in cybersecurity is not the only pathway for entry into the field, and work experience is more highly valued by some employers than a degree, the majority of people working in this field need degree-level education to perform at even basic levels of effectiveness.

This analysis helps us understand that the role of universities in cybersecurity workforce development may be somewhat less important than, or at least different from what the Chinese government might imagine. In July 2016, the Central Leading Group on Cybersecurity and Informatization approved "Suggestions on Strengthening the Construction of Cybersecurity Discipline and Personnel Training" (CAC 2016) that were subsequently attributed to Xi Jinping. Recognising that the talent gap is "huge" and that the "talent structure is not rational", the document outlined eight measures:

- Speed up investment in research and building of laboratories
- Transform delivery mechanisms by having universities deliver professional training courses, and by bringing in specialists from think tanks to help the transformation
- Create reasonable cybersecurity text books that meet the national requirements
- Build strong teams of educators by relying more on experts, internationalising the teaching staff, especially by employing high-end foreign talents
- Encourage enterprises to participate in the policy setting and training of cybersecurity talents in universities
- Strengthen on-the-job training for the cybersecurity employees and establish classifications and competency standards for cybersecurity jobs
- Strengthen cybersecurity awareness and skills training for the public through full use of the Internet, radio, film and television, newspapers and magazines and other platforms, especially educational curricula
- Improve the economic incentives for cybersecurity talent cultivation in order to promote internationally competitive and influential talents.

These measures were calling for root and branch renovation of cybersecurity education more than two years after Xi had called in February 2014 for the country to start producing high quality talent teams. While radical in tone, the persistence of the traditional Chinese education approach of putting everything in approved text books (that can take years to get approved) is one that appears somewhat inappropriate in this case and would hold the other reform objectives back. In December 2016, when China issued its national cybersecurity strategy, these proposals for talent development resurfaced in more detail. In August 2017, CAC issued "Administrative measures for demonstration projects in establishing first class cybersecurity colleges" (CAC and MoE 2017).

High-end talents are scarce. In terms of educational status, people with the undergraduate degree take up 61.8%, master degree or above account for about 9.6%; diploma occupy 25.2%; other about 3.4% (Zhang et al. 2016: 5). People who have studied both technology and management are even more scarce than high end talents.

Graduates are seen as not meeting employer expectations. And there is a geographical imbalance in talents, with most located in the eastern part of the country.

One Chinese specialist, relying on survey data, has developed a framework for education in the field based on post competencies (Sun 2017). The author identified four sets of competencies in information security: theoretical research, technological development, management services, and education and training (447). Within management services, the posts in most demand were engineers in security testing, security services, data recovery and network security. Of some interest, the priority competencies were not technical. They were in order:

1. Legal literacy and ethics
2. Autonomous learning ability
3. Ability to cooperate with others
4. Emotion management in a variety of circumstances
5. Responsibility consciousness
6. Communication ability
7. Professional knowledge
8. Professional skills.

In 2017, China's most popular employment advertising website, Zhaopin.com, teamed up with Qihoo 360, to produce a report on the market in China for cybersecurity talents (Zhaopin.com and Qihoo 360 2017). While the report has its limitations, the results on data from the first half of 2017 are worthy of further refection and analysis. In broad terms, the report points to a red-hot but highly distorted market where the majority of advertisements for cyber security posts require no qualifications in the field, though most demand at least a tertiary degree in any field (1). The growth rate in jobs advertised was 232% year on year, but only 11% of applicants have qualifications in cybersecurity. The geographic focal points are noteworthy, with Beijing, Shanghai, Shenzhen, Chengdu and Guangzhou accounting for just over 50% of all vacancies. In both Beijing and Shanghai, as in several other big cities, demand outstrips supply, but in Shenzhen there is an oversupply of applicants (2). The biggest shortfall in the supply is in two-year college graduates (only 60% of jobs advertised at that level are filled), with university-trained people representing a surplus (3). The majority of jobs are technical, with only 4.6% of jobs advertised in managerial roles (9–10).

The sectoral focus is unusual, with the government and public institutions only accounting for 4.8% of demand (7–8). The biggest areas of demand include big data and emergency response (16–20). The report concludes that the shortage of cyber security talents will haunt China for the long term.

China's intention to move aggressively outside the university sector to expand education opportunities in the field was demonstrated in August 2017 when it announced the start of construction for an Information Security Institute in Wuhan (Hubei province) that would open in 2019 and provide non-degree training for 10,000 students (over an unspecified time period) at a cost of US$751 million (Xinhua 2017).

2.9 Conclusion

China faces many of the same dilemmas of cybersecurity education policy and delivery that other countries do:

- the field is rapidly changing as technologies shift
- the core body of knowledge for university delivery is heavily dominated by engineering, maths and computer science
- little attention is paid to social science aspects
- there are only weakly developed capabilities and options for students to conduct even medium complexity simulations and experiments.

At the same time, China must deal with severe constraints of its own making. These include:

- a rigid and authoritarian university system that resembles the one created in the Soviet Union between 1951 and 1977 (Zhang 2017: 150)
- only low levels of internationalization
- intrusion of CCP supervision that affects student life and academic merit for teaching and research staff.

While some Chinese research institutions are achieving impressive gains in more technical aspects of the sciences of information security, they appear to have only a handful of avenues for evolution of research and teaching around social science aspects of the field: management, privacy, economic impacts, and politics.

In general, there remains a huge deficit in China in the numbers of people trained or educated in information security to match the ever growing demand for them. While this is a familiar story in all countries, the sheer size of the user population in China (individuals, business, and government agencies) and the government's insistence on a mass cyber surveillance system for internal security create heavy pressure on work force development that few countries must deal with. Market forces are in play in the labour supply dynamic in China, but unfortunately for it, the market for the best skills is a global one. Since the major foreign consumer of Chinese information security talent is the United States or corporations based there, and since there is little systematic collection of data on the numbers and talents of Chinese being taken up by that market, we are not in a position to reliably assess its effects on the supply of the best talents in China.

We do know that Chinese citizens with first degrees from the country's universities fill leadership roles in cutting edge technological research for information security in foreign institutions, such as MIT and Imperial College London. We also know anecdotally that many of these people see themselves irrevocably as participants in a globalized system of knowledge, owing no allegiance to any techno-nationalist vision or mass surveillance ideology of the Chinese state. Some Chinese expatriate specialists actively work to defeat the government's technologies.

What China can assert with confidence, as this chapter suggests, is that beginning in 2015, the education system began a journey toward reform and opening up in the field of information security somewhat akin to the journey began by the economy as a whole in 1979. China is not four decades behind, but it has a long hard race ahead to begin to catch up to the education and research standards of the country its leaders aspire to emulate in this field: the United States. Subsequent chapters will add nuance and contours to this broad judgement, showing increasing visibility of pockets of excellence and internationalization in Chinese studies of, and mastery in practice, of information security.

References

ATIP (2015) ATIP: cybersecurity research in China. Alberquerque NM. Sponsored by the United States government. www.dtic.mil/get-tr-doc/pdf?AD=ADA619229
Austin G (2014) Cyber policy in China. Polity, Cambridge UK
Austin G (2017) Human capital for cyber security: the Australian case. Briefing Paper #2. Australian Centre for Cyber Security. University of New South Wales Canberra. https://unsw.adfa.edu.au/australian-centre-for-cyber-security/sites/accs/files/uploads/Briefing%20Paper%232.pdf
Austin G, Zu H (2018) Cyber security education in China: An overview of selected universities. Discussion Brief #4. Australian Centre for Cyber Security. University of New South Wales
CAC (2016) Suggestions on strengthening the construction of cyber security discipline and personnel training. Cyberspace Administration of China. In Chinese. 8 July 2016. http://www.most.gov.cn/tztg/201607/t20160708_126464.htm
CAC and MoE (2017) Management method of construction and demonstration project of first-class cyber security institutions. In Chinese. Issued by the Secretariat of Office of the Central Leading Group for Cyberspace affairs and the Ministry of Education 8 Aug 2017. http://www.moe.gov.cn/srcsite/A16/s3342/201708/t20170815_311176.html
Chen H, Maynard SB, Ahmad A (2013) A comparison of information security curricula in China and the USA. Edith Cowan University Research Online. http://ro.ecu.edu.au/cgi/viewcontent.cgi?article-1152&context+ism
ComputerworldHK (2009) ISACA expands into China. Computerworld.com. 24 Nov 2009. https://www.cw.com.hk/it-leadership/isaca-expands-into-china
Conklin W, Cline R, Roosa T (2014) Re-engineering cybersecurity education in the US: an analysis of the critical factors. 2014 47th Hawaii International Conference on System Sciences. http://ieeexplore.ieee.org/stamp/stamp.jsp?reload=true&arnumber=6758852
CUAA (2017) Alumni Association 2017 China University new discipline rankings. Chinese University Alumni Association. 18 Aug 2017. www.cuaa.net/paihang/news/news.jsp?information_id=133226
Feng D, Ling D, Yung M (eds) (2005) Information security and cryptology: first SKLOIS conference, CISC 2005, Beijing, China, 15–17 Dec 2005. Proceedings. Springer Science & Business Media, Berlin
Hartson HR (1983) Teaching protection in computing: a research-oriented graduate course. Comput Secur 2(3):248–255
Hong Yu (2017) Reading the 13th five-year plan: reflections on China's ICT policy. Int J Commun 11(2017):1755–1774
Hui Z, Tan Q (2016) Cyberspace security in the era of data economy: global and chinese contexts. In: Hui Z, Tan Q (eds) Annual report on development of cyberspace security in China. Social Sciences Academy Press. Blue Book, Beijing. pp 1–16 (In Chinese)
Kim D, Song Q, Liu J, Liu Q, Grimm A (2017) Building world class universities in China: exploring faculty's perceptions, interpretations of and struggles with global forces in higher education. Compare: J Comp Int Educ 1–18

Lü W (2017) Data on China's PhD completions related to cyber security. Briefing paper #3. Australian Centre for Cyber Security. University of New South Wales Canberra. https://unsw.adfa.edu.au/australian-centre-for-cyber-security/sites/accs/files/uploads/Briefing%20Paper%233%20Final%2023%20NOV.pdf

Luo Y (2013) Building world-class universities in China. Institutionalization of world-class university in global competition, pp 165–183. Springer, Netherlands

Luo Y (2011) Study on the current situation of information security and countermeasures in China. Energy Procedia 5:392–396. http://www.sciencedirect.com/science/article/pii/S1876610211010034

MoE (2016a) Number of students in higher education institutions 2015. http://en.moe.gov.cn/Resources/Statistics/edu_stat_2015/2015_en01/201610/t20161011_284371.html

MoE (2016b) The academic degrees committee of the state council agrees to add cyberspace security. Notice of doctoral degree authorization at level 1. Decree [2016] No. 1. In Chinese. 28 Jan 2016. http://www.moe.gov.cn/s78/A22/A22_gggs/A22_sjhj/201603/t20160304_231944.html

NSF (2015) Doctorate recipients from U.S. Universities: 2014. National Science Foundation, National Center for Science and Engineering Statistics. https://www.nsf.gov/statistics/2016/nsf16300/digest/nsf16300.pdf

NSF (2017) Doctorate recipients from U.S. Universities: 2015. National Science Foundation, National Center for Science and Engineering Statistics. https://www.nsf.gov/statistics/2017/nsf17306/data/tab26.pdf

OECD (2016) Education in China. A snapshot. organization for economic cooperation and development. https://www.oecd.org/china/Education-in-China-a-snapshot.pdf

Olijnyk NV (2014) Information security: a scientometric study of the profile, structure, and dynamics of an emerging scholarly specialty. Dissertation. Long Island University, CW Post Center

Popkin K (2017) China's quantum satellite achieves 'spooky action' at record distance. Sci Mag. 15 June 2017. http://www.sciencemag.org/news/2017/06/china-s-quantum-satellite-achieves-spooky-action-record-distance

Ren J, et al (2017) Ground-to-satellite quantum teleportation. arXiv. https://arxiv.org/ftp/arxiv/papers/1707/1707.00934.pdf

Schneider FB (2013) Cybersecurity education in universities. IEEE Secur Priv 11(4)

Shuang Y, Gao W, Pang M (2016) A comparative study on the highly cited papers of the top eight engineering universities. COLLNET J Scientometrics Inf Manage 10(2):311–320

Sun Q (2017) Research on the curriculum system of information security and management specialty in higher vocational education based on post competency model. DEStech Transac Social Sci, Educ Human Sci

Tech Review (2017) First object teleported from earth to orbit. MIT Technol Rev. 10 July 2017. https://www.technologyreview.com/s/608252/first-object-teleported-from-earth-to-orbit/

Wang X, Lu W, Zhou X, Guo W (2017) Analysis and thoughts on the development of the information security profession in advanced education. In: Humanity and social science: proceedings of the international conference on humanity and social science (ICHSS2016), pp 84–95

White House (2017) Presidential executive order on strengthening the cybersecurity of federal networks and critical infrastructure. Office of the Press Secretary. 11 May 2017. https://www.whitehouse.gov/the-press-office/2017/05/11/presidential-executive-order-strengthening-cybersecurity-federal

Xinhua (2014) Network and information security development, personnel construction is the key. In Chinese. 27 Nov 2014. http://news.xinhuanet.com/politics/2014-11/29/c_1113456181.htm

Xinhua (2017) China starts building internet security institute. 23 Aug 2017. http://news.xinhuanet.com/english/2017-08/23/c_136549729.htm

Zhang H, Han W, Lai X, Lin D, Ma J, Li J (2015) Survey on cyberspace security. Sci China Inf Sci 58(11):1–43

Zhang H, Yu H, Zhai J, Yu X (2016) Suggestions on cyber security talents cultivation. Chin J Netw Inf Secur 2(3):1–9. (In Chinese)

Zhang L, Wu Y, Yang G (2017) M&S as a profession and discipline in China. In: Tolk A, Ören T (eds) The profession of modeling and simulation: discipline, ethics, education, vocation, societies, and economics. Wiley, New York, pp 167–182

Zhang W (2017) The path of China higher education governance reform based on comparative study on the governance pattern of higher education in the United States and Japan. In: Humanity and social science: proceedings of the international conference on humanity and social science (ICHSS2016), pp 150–155

Zhaopin.com and 360 Cybersecurity Center (2017) Research report on cybersecurity talent market in China. In Chinese. Beijing. Available at: http://zt.360.cn/1101061855.php?dtid=1101062370&did=490711371

Chapter 3
Chinese Views of the Cyber Industrial Complex

Abstract This chapter outlines the government's view of national policy for S&T and industrial development in the field of cybersecurity. It draws on a roadmap for S&T development developed in 2011 by the Chinese Academy of Sciences. It documents vigorous leadership engagement in these issues beginning in 2016, and their policy imperative of indigenization. It discussed the emergence of the private sector in cybersecurity relative to the better established state-owned sector.

Chinese leaders accept that cybersecurity has an inevitably globalized character. At the same time, they have expressed an urgent need for China to develop its own cyber industrial complex. Without using this precise term, the Chinese government has promoted a view that it must construct such a system at home if it is to protect national interests both in peacetime and against the contingency of war. China made history in its 13th Five Year Plan released in 2016 when it elevated national cyberspace security to be one of only six high priority development areas in science and engineering, along with quantum computing and communications, and four others. It is the first country to give cyberspace security such centrality in national science and technology policy. This was the first five-year plan to include a separate headed section on cybersecurity (Yu 2017: 1755).

Because of the heat and political primacy of the techno-nationalist rhetoric around the cybersecurity industry, the room for serious discussion of several key issues in this sector by Chinese scholars is reduced if not obliterated. These issues that are under-analysed in Chinese and non-Chinese sources in respect of China include:

- The balance to be struck between self-referential state interests and the interests of individual citizens (because citizens experience cyber harm differently from the state)
- The balance to be struck between self-referential state interests and the interests of commercial enterprises (because firms experience cyber harm differently from the state)
- The balance to be struck between national cyber industrial capabilities and a globalized ICT economy, based on mobile labour, mobile capital and highly fluid and evaporating political loyalties.

G. Austin, *Cybersecurity in China*, SpringerBriefs in Cybersecurity,
https://doi.org/10.1007/978-3-319-68436-9_3

This chapter looks primarily at the last of these questions. It relies on a rich fund of comments in the Chinese media and on conversations between the author and leading Chinese figures in the sector. It also relies on a number of world class studies by scholars, Chinese and non-Chinese, on these issues. The focus is not just on industry policy for the sector but also on the realities of globalization. What should the cybersecurity industry for China look like? Where is the necessary boundary for China between sovereign capability for certain national security needs and active participation in a non-national global economy? How does a globalized economy interact with China's national innovation system? As Lindsay (2017) asks so well of the global scene (and as this chapter asks for China): what is the international political economy of cybersecurity? The chapter is informed by a research note on select Chinese and foreign cyber security companies operating in China prepared in connection with this book project (Austin and Meng 2018).

3.1   S&T Roadmap for Cybersecurity

Of all the G20 members, the Chinese government is probably the one most committed to leading the transformation of its national economy and society through informatization—the exploitation of advanced information and communications technologies in all walks of life. China has a long way to go to catch up in most measures of cyber power. But it is moving as rapidly as it can. For example, in 2012, Chinese scientists claimed to be the first in the world to have succeeded in the physical teleportation of the electronic properties of remote sub-atomic particles, one of the important foundations for quantum communications (Bao et al. 2012). By 2017, as mentioned above, Chinese scientists reported the first teleportation of an object into orbit by replicating the 2012 experiment to teleport the electronic properties of remote sub-atomic particles between earth and outer space. This was followed by a practical demonstration of quantum key distribution during a video conference call using China's dedicated quantum communications satellite (Nordum 2017).

It seems to be the case that the government of China had decided at least by 2010 on the indigenization policies in the cybersecurity sector that have become so prominent under Xi Jinping. Policy making in China on such matters typically moves very slowly and conservatively, so it is unlikely that the policy directions for S&T development had matured through Xi's leadership sufficiently after he took over as General Secretary of the CCP in November 2012 to be credited to him.

The scale of the China's ambition to become a world leader in the S&T base of cyber power is documented in a 2011 plan by the Chinese Academy of Sciences (CAS) called *Information Science and Technology in China: A Roadmap to 2050* (Li 2011). The vision is staggeringly ambitious and complex. It sees China approaching the frontiers of science, economics, and social organization in the sphere of information technology by mid-century (Austin 2014: 110). At the time, Chinese leaders were already advancing the idea that the country needed to do more to boost the quality and market share of the cybersecurity industry.

One impetus for the 2011 report was a 2009 strategy document, *Technological Revolution and China's Future: Innovation 2050*, from the Academy of Sciences (Lu Y 2010), which served not just as an overarching mobilizing document, but also marked the launch of a series of 17 subsequent sector-based roadmap reports also looking ahead to 2050. The 2009 foundation report on innovation, which had involved some 300 Academy researchers and experts for more than a year, recommended that China prepare itself for a new revolution in S&T in the coming ten to 20 years in green energy, artificial intelligence, sustainable development, information networking systems, environmental preservation, space and ocean systems, and, most interestingly, national security and public security systems.

The 2011 IT sector roadmap had a chapter devoted to information security under the heading "Establishing a Technical System for National and Social Information Security" (Li 2011: 133–142). The roadmap set targets for the short, medium and long term as set out in Box 3.1. In essence, the roadmap foresees three stages: integration of cybersecurity services and technologies running to 2020; a second phase over 15 years of automated cybersecurity services; and a third phase running from 2035 to 2050, in which quantum cryptography and advanced artificial intelligence play the dominant role ("smart cybersecurity") (Li 2011: 133–34).

Box 3.1: CAS Roadmap for Information Security Technology

Short-term targets (2010–2020): Construct a network-based lightweight information security foundation technological system. Research the establishment of a large-scale and collaborative cyberspace security technological system. Research and develop an integrated security service technological system. Achieve the commercialization of point-to-point optical fiber quantum cryptography communication systems and the trial success of a 70 km metro optical network quantum key distribution.
Mid-term targets (2021–2035): Build the base technological system for automated and intelligent information security. Research the establishment of a highly credible and interactive cyberspace security technological system. Research and develop the technological system for automated information security services. Achieve the commercialization of metro quantum cryptography communication systems, and expand Metropolitan Area Networks to Inter-metropolitan Networks.
Long-term targets (2036–2050): Build up a dynamic and high-intensity based information security technological system. Research the establishment of high availability and self-organized cyberspace security technological systems. Research and develop intelligent security service technological systems. Implement global practical secure communication networks based on quantum key distribution.

The CAS vision of the long-term trend was profound. Cybersecurity would no longer be about computing security "this kind of security is either reduced or disappears entirely" (134). There would be a need to research information security technology alongside "unconditional security (that is, having nothing to do with the computing power)" and aiming for "technology that can adapt to different environments and modes (134). The day to day needs would have to coexist with new advanced research. But the foundation of all security in cyberspace would be the high intensity design of cryptographic systems and analysis (135).

Ten priorities for technology research were identified (135–37):

1. purification technology (high speed content control, tracing and purging)
2. active immunization (automated identification, isolation, and rejection of threats)
3. a social credit system for users based on audits, assessments and documentation of user conduct/misconduct
4. supervision and management of monitoring control systems
5. unified certificate authorities and audit techniques for multi-network integration (separation of business systems from their security, including through the implementation of access control of "dynamically complex fine-grained functions")
6. infrastructure components of information security (highly-efficient, highly-reliable and intelligent)
7. detection and early-warning technologies for malicious code
8. collaborative protection of large-scale network systems (correlation analysis for detection of malicious code, automatic generation of protection strategies, and coordinated response)
9. secure storage of and access to super large data (including intrusion tolerance mechanisms and integrity mechanisms)
10. trusted computing environments (platform security architecture and hardware).

Other high level priorities in the roadmap included technical systems for information security assessment (138–40); and a new communications network with security based on quantum cryptography (140–42).

3.2 Government Vision

The roadmap was directed at state-run institutions, such as government research institutes, universities and SOEs, such as the China Electronics Technology Group Corporation (CETC). The domestic privately-owned cybersecurity sector was miniscule in 2011. The timing of the release was in some respects not that auspicious. It occurred about a year out from the leadership changeover in the CCP due in November 2012 as the ten-year term of General Secretary Hu Jintao was drawing to an end. Normally, this twelve-month period just before the changeover is one of policy stagnation. On this occasion, it was moreover a period of considerable leadership turmoil. In addition, such blueprints are more aspirational rather than likely to be implemented in full. Headline policies in selected areas of industry R&D policy are likely to be promoted or ignored according to the political priorities of the day, while much of the more detailed elements would normally be ignored. Added to this is the consideration that the IT roadmap was, as mentioned, one of 17 similar strategies from the CAS for other sectors.

The government vision in implementing the roadmap and the response of the institutions intended to execute it have been shaped by what the document had called "day to day needs" and these emerged with a vengeance in 2012 and 2013: exposure of the leaders to cyber espionage by elements of their own internal security agencies, exposure by the Mandiant report (2013) of a key PLA unit (61398) as almost totally inept at its own cybersecurity, and exposure by Edward Snowden of immense cyber vulnerabilities in China. In response, the leadership announced in February 2014 (as mentioned earlier) that they would do everything needed to make China a cyber power. By 2016, the leaders were taking a more aggressive stance toward what this meant in both in basic science and in industry policy.

In a speech at the National Meeting on Cybersecurity and Informatization on 19 April 2016, Xi (2016) addressed the domestic industrial policy aspects of the cyber power strategy, of which security was one of six underpinnings. The others were: embracing the internet, using the internet for the expression of public opinions; breakthroughs in core Internet technologies; working with the private sector; and IT talents. In this he was applying to the cybersecurity industry key judgements he was to make in June 2016 about China's S&T base in general, as mentioned below.

On core technologies. Xi boasted in the April speech that "four of the world's top 10 Internet companies are Chinese" but observed "that there are quite a number of areas in which we are falling well short of leading world standards". He said "this gap is biggest in terms of core Internet technologies" and reliance on others for these "is our biggest threat". Even then he limited his vision to overtaking "in certain fields and areas". To achieve this, he called for determination and organization: "strategic plans for the development of core technologies and equipment in the information domain; draw up roadmaps, schedules, and task briefs; define short-, medium-, and long-term goals; respect objective laws by advancing initiatives one level, category, and stage at a time; and ensure that once we have started we never let up". And he called for focus": "take aim at the cutting edge ... while bearing in mind China's national conditions; enhance our planning in key areas and links with a view to assuming the strategic high ground".

He then defined core technology by reference to three categories: "basic and generic technologies; asymmetric and "trump-card" technologies; and cutting edge and revolutionary technologies". He claimed that in all of these areas, China is "on a level playing field with other countries", a claim repudiated quite explicitly by Chinese officials in many places (as documented in this book). The first step, Xi said, involved striking the "right balance between openness and autonomy". Observing that the Internet cannot ever be closed off and such a policy was not the answer for China, he said the country needs to "encourage our Internet and IT companies to go global, to deepen international exchanges and cooperation, and to participate actively in the Belt and Road Initiative", while welcoming all foreign companies to China "as long as they comply with our laws and regulations".

He said China needed to take control of core technologies "through self-innovation, self-reliance, and self-improvement" and from the country's own R&D. This had to be combined with an "open approach to innovation", since it would only be through direct competition with a "strong adversary" that China could avoid complacency and understand how far the country is lagging. He said the country had to have a view on "which technologies can be introduced from abroad, provided they are secure and controllable; which technologies can be introduced for the purpose of assimilation, absorption, and re-innovation; which technologies can be developed in collaboration with other parties; and which technologies must be developed independently through our own innovation". He said none of this would be possible without "good basic research" though the results so far of a lot of expenditure on this "have not been particularly outstanding". He called for a robust policy of developing new products from basic research: "If a core technology becomes detached from its industrial chain, value chain, and ecosystem, failing to link upstream and downstream segments of the chain, its development may very well have been a wasted effort".

He lamented the lack of commercial collaborations between Chinese corporations, citing as a model "the way that Microsoft, Intel, Google, and Apple do". He said it was high time "to free ourselves from the constraints of departmental interests and faction-like bias" and "overcome petty, small-minded thinking and form synergy". He called for more alliances between universities and firms, between SOEs and privately owned firms, and innovative approaches to venture capital: "more tightly-knit capital-based collaboration mechanisms, establishing investment companies for the R&D of core technologies".

In June 2016, Xi characterized China's current status in science and technology (not just information-related fields) as weak: "The situation that our country is under others' control in core technologies of key fields has not changed fundamentally, and the country's S&T foundation remains weak" (Xinhua 2016a). He said that S&T is the bedrock of the country's power, and that "Great scientific and technological capacity is a must for China to be strong" (Xinhua 2016b). Xi publicly endorsed a formula for what the transition to 2050 looks like. China should aim to become one of the most innovative countries by 2020 and a leading innovator by 2030 before realizing the objective of becoming a world-leading S&T power by the centenary anniversary of the founding of the People's Republic of China in 2049 (Xinhua 2016c). Xi made more general invocations in connection with the meeting in April 2016 of the Leading Group on Informatization and Cybersecurity.

Soon after, when the World Economic Forum published its annual "Network Readiness Index" in 2016 comparing the cyber capability of 143 countries, China was sitting in 59th place, down from 36th in 2011; and lower than countries like Montenegro, Mauritius and Turkey (WEF 2011, 2016). China's position arises in part because of the per capita comparisons in the index which distort the relative wealth and technological capabilities of countries. China has, of course, made great strides, but the United States, Japan, Korea, Singapore and Europe have raced ahead as well. As noted by *China 2030* (DRC and World Bank: 2013: 17), "innovation at the technology frontier is quite different in nature from simply catching up technologically".

In July 2016, the government translated these general ambitions for S&T announced by Xi in June into goals specifically for the cyber power ambition:

2020: "key technologies achieve an advanced international level, the competitiveness of information industry upgrades significantly on a global scale, and informatization becomes the leading force of modernization"

2025: "establish an internationally advanced mobile communication network ... and rid the country of its dependence on foreign key technologies"; "aim for "well-developed advanced technology industries, leading applications and infallible cybersecurity; with major transnational internet information enterprises and competitiveness taking shape".

2050: "informatization will play a pivotal role in establishing a prosperous, democratic, civilized and harmonious modern socialist country"; be a cyber power and a "leader in the development of informatization around the globe" (Zhang 2016).

The government also released six sets of policy guidance that had been in preparation since 2014 on the following broad subjects (Zhang 2016):

1. "coordinated promotion of IT technologies in central and local governments, and between the Party, government and military.... coordinating the roles of the market and government, focusing on interim and long-term targets, and major issues of informatization in various fields"
2. development of core technologies
3. boosting informatization "in economics, politics, culture, society, and also ecology, national defence and the military"
4. building a people-oriented approach to implementation of the strategy
5. highlighting the role of informatization in international affairs and strengthening China's influence on the global cyber stage
6. ensuring national cybersecurity.

For the current leaders, one of the primary goals of policy is to rid China of foreign "control" over core technologies. They see the major pathway to this goal as the rapid consolidation of the domestic ICT sector and its projection on the global stage. They are also very anxious to overcome their technological inferiority through greater indigenization of cyber-related industries, even if this strategy runs up against world trade norms, as many governments and corporate leaders have argued and as China itself has been forced to acknowledge on occasion.

Xi's speeches seemed to contrast in part with the mentality behind a speech in 2016 by the chief economist of the China Electronics Information Industry Group (known in English by the acronym CEC for an earlier incarnation, China Electronics Corporation), Wu Qun. In the annual meeting of military industry in 2016 (Wu 2016), Wu identified three phases of China's development in the cybersecurity sector, which ran in parallel with the trajectory of CEC: electronic country, information country, and-still to come-cyber power.

Wu implies that his company responded to the CAS roadmap. He noted that in 2011, CEC "decided to build a new development strategy for the group company with its own controllable network information system engineering, and began to explore and explore to build a new strategic capability". He cited "national strategy and historic needs", saying it is the "responsibility of CEC as a central enterprise to safeguard national security". CEC sits alongside another SOE, CETC, as the two most important cybersecurity companies in China, both with a heavy military and national security history. The Chinese press has labelled these two companies as the "vanguard of national information security" (china.com 2015).

Nevertheless, as Wu observed, cybersecurity "has not constituted an industrial system in China", and its security products, such as anti-virus software and firewall technologies "can only be taken as fragments". In contrast, he said that CEC is trying to set up an "information security industry system" by focusing on three areas:

1. Dominate the production of core products of basic industries in China and build an "independent controllable industrial supply chain";
2. Put safety supervision and maintenance in center stage, and build a controllable security service system;
3. See industrial security in the broad as part of the process of digital industrial control systems.

Wu reported that "after four years" (the period since the 2011 decision), CEC has more or less "achieved a complete and controllable independent platform and hardware products, and formed a complete localization ecological chain".

3.3 Industry Trends

The main distinction to be made in this sector of the corporate world in China is between the security firms working under close government remit (that is, the state-owned enterprises or SOEs), and those operating with a greater degree of separation (that is the private sector). With policy and legal reforms introduced by China since 2014 to strengthen the security standards and practices of Chinese firms and to promote the domestic industry, the freedom of maneuver of those private firms, once more distant from government, has been reduced. On the other hand, they are continuing to benefit from a surge in corporate and government spending on cybersecurity, bolstered by the priority accorded the sector in the 13th Five Year Plan.

Citing MIIT, Hui and Tan (2016:188) reported the share of the country's ICT industry taken by security in 2014 was only 0.28%. While this represents a doubling of the share in 2010, the authors saw this an alarmingly low share that revealed the dangerously low priority accorded to this mission throughout China. This low share in 2014 compares, according to Chinese sources, with an estimated 14% in some Western countries.

Four sectors dominated the cybersecurity market in China in 2014: government, finance, telecommunication and energy, which together accounted for 60% by market value (iiMedia 2017). Product sales had a huge dominance over services: around 71 and 29% respectively. Main product lines included firewall, identity management (fingerprint and iris recognition), access control, unified threat management, and secure content management. Within the product lines, hardware dominated over software (50% compared with 21%). The services most in demand were security assessment and security integration, accounting for 90% of the services market (Hui and Tan 2016: 188). In the entire services market, training of personnel accounted for only 0.75% of outlays.

According to data from the China Academy of Telecommunication Research (CAICT), under MIIT, the size of the domestic information security industry in 2014 (state-owned and privately owned) was about 39.37 bn RMB (cited by Hui and Tan 2016: 188). This was not too different (on an unadjusted exchange rate basis) from the annual global turnover in 2014 of the single leading U.S. corporations, Symantec, which was US$6.5 billion (Symantec 2015: 1). CAICT reported that in 2014 the output of more than 20 listed private sector companies in China dedicated to cybersecurity was 9.714 bn RMB.

After 2014, there have been major changes, with new investment flowing into the sector, many more firms established, including 30 that successfully listed as public companies, and with consulting companies beginning to undertake serious and sustained research on the industry and the security situation in China for the first time (Hui and Tan 2016: 188). By 2015, the sector in China was registering a growth rate of 15% annually. According to a leading securities firm, reporting on the private sector in 2015: "China's information security technology and construction level is relatively backward, and the security situation it faces is more severe. The importance of information security has been elevated to unprecedented heights" (Huatai Securities 2015: 1). In contrast, a report by the Security Research Institute of the China Academy of Information and Communications Technology (CAICT 2016: 194–96) concluded that "in recent years, the strength of domestic security enterprises has steadily improved".

In 2017, Sinolink Securities released a report that speaks to China's views and progress on creating a cyber industrial complex, though not using that term (Sinolink 2017). The report noted that on key indicators, the domestic industry was roughly where U.S. sector was in 2001. The report expected that the establishment of the Central Leading Group on Cybersecurity and Informatization, announced in February 2014, and implementation of the law on network security would allow the domestic security industry to replicate the decade long golden era in the U.S. sector after 2001. During these ten years in the United States, cybersecurity investment grew from 2% of the ICT sector to 11%. The report projected more than 25% growth per year in the sector in China.

If the CAS roadmap for IT was released at an inauspicious time, it needed only a few years to find a much higher relevance after the 2014 cyber power announcement. By December 2014, the government introduced new regulations for intended to help promote the rapid growth of China's domestic industry. In May 2015, the country issued a new Military Strategy enshrining the idea that cyberspace along with outer space have "become new commanding heights in strategic competition among all parties" (SCIO 2015). The same month, the National People's Congress released a draft bill on National Security (passed in July) that gave a special place to cybersecurity in its provisions for strengthening government control over foreign technologies and related investment in China. (The text of the draft legislation and laws of China can be found at www.chinalawtranslate.com.) Also in July 2015, China released a new draft law on cybersecurity with sweeping provisions on control of foreign technologies and data management. The bill was passed and entered into force on 1 June 2017. At that time, the government rushed through deliberation on a new Law on National Intelligence. Other related laws passed and entering force since the cyber power announcement included a Counter-Espionage Law (2014) and a Counter-Terrorism Law (2016). Thus, the vision of cyber power appears to be one that prioritises consolidation of the technological and industrial attributes of cyber power and the application of such power to keep the CCP in control at home. The vision therefore has techno-nationalist and authoritarian dimensions.

In this light, we might assess many of the developments in the recent past as ominous and presaging some sort of confrontation between China, with its increasing cyber power, and the West, with its cyber power imagined to be either static or in relative decline. This reading however would not represent the totality of the Chinese leadership's position on the balance to be struck between China's aspirations for sovereign capabilities and a globalized cyber industrial complex. On the one hand, even a superficial reading of new Chinese laws and draft bills impacting on cyber policy, reveals unusual and positive attention to the importance of international economic relations. Chinese leaders know, as leading specialists in China and elsewhere recognise, that the country's cyber sector will remain highly dependent on technology transfer and investment from the most advanced countries for many years, probably decades. On the other hand, in spite of reports of a blacklist of U.S.-based companies that have been named publicly by Snowden as participating in cyber espionage against China (Pasick 2015), the country continues to solicit and accept massive investments from them. For example, in June 2015, Cisco's incoming CEO, Chuck Robbins, announced in a visit to China that the company would invest US$10 bn in China (Patton 2015), a move widely seen as an effort to rebuild trust with China in the wake of the Snowden revelations.

China's formal position is that the foundations (the attributes) of its own cyber power can only be built through international collaboration, not just with the 60 or so like-minded countries, such as Russia, who support Chinese positions in international forums, but also with the technology-leading countries, like the United States and Japan, and their partners in Europe. This understanding helped drive China by 2015, after a decade of multilateral effort by many states including China, to sign up (unofficially) to "possible voluntary" norms elaborated by the UN Group of Governmental Experts on cooperative relations in security aspects of cyberspace (Marks 2015). The political weight we can attach to China's support for diplomacy as a foundational element of cyber power is reflected more convincingly in a formal agreement with Russia in 2015, in which the two countries agree not to undertake actions like "unlawful use or unsanctioned interference in the information resources of the other side, particularly through computer attack" (Russian Federation 2015).

Thus, at the outset of the pursuit of China's ambition to accumulate the attributes of cyber power there is a fundamental conflict or contradiction. In classic chauvinistic fashion, China wants to rid itself of foreign technological control but says it needs to participate in a global system of the exercise of cyber power to even begin the journey. Moreover, President Xi has specifically identified the need for China and the United States to work together to avoid the "Thucydides trap", the propensity for conflict between an emergent power and an established power because of hubris or fear (insecurity).

There are numerous non-Chinese assessments agreeing with this proposition about China's acceptance of the need to "collaborate to compete", but the sentiment is well captured by a U.S. study: "China's S&T investment strategy is ambitious and well-financed but highly dependent on foreign inputs and investments. Many of its stated S&T and modernization goals will be unachievable without continued access to and exploitation of the global marketplace for several more decades" NRC (2010: 23). This study went on to say that "Within the IT sector, absent a sharp change in course, China's stance toward technology acquisition is likely to evolve into a self-inflicted wound of substantial dimensions. ... China risks being 'designed out' of the increasingly internationalized cutting-edge IT R&D environment, a process that has already begun" (27).

3.4 Command Economy

The government now operates a controlled structure for cybersecurity vendors to obtain a standing security clearance, though the exact start date of this security vetting process is not clear. The system is different from the more common cyber security certification for government procurement (China Certification of Information Security), which is currently held by over 90 domestic firms and for which foreign firms were permitted to register beginning in 2015. The primary beneficiaries of the more restrictive higher-level national security vendor certification in recent years include two state-owned enterprises (SOEs) and several privately-owned firms. The cybersecurity SOEs (Chinasoft and China Cybersecurity) described in Austin and Meng (2018) are in fact subsidiaries of separate large state-owned conglomerates mentioned earlier, CEC and CETC. The private firms include NSFocus, VenusTech and KnownSec. These private companies are "background-controlled": specifically certified as 100% Chinese by capital investment and by personnel, with the personnel review almost certainly including some assessment of their competency as well as loyalty. This type of firm (six private firms in total according to one source) are the only ones which qualify, in normal circumstances, for higher-level government procurement contracts in cybersecurity. Another company outside the group, Antiy, is close to the government and appears to enjoy a similar standing. The SOEs and these favoured companies transfer a large number of their government-assigned projects to contractors or to shell companies operating under their official incorporated business holding.

In line with the global scene however, the cybersecurity business in China has become an important activity of large firms in the ICT or e-commerce sectors whose primary business was not initially cybersecurity. In China, such firms include Alibaba, TenCent and Baidu, all of which have divisions or subsidiaries selling security products and/or services.

Before the government imposed this structure of having a small number of background-controlled firms, the business environment in information security was highly corrupted. According to well-placed Chinese specialist, many companies which received government contracts in information security were simply pirating tools and architectures from foreign vendors, often only changing a small amount of the code. In 2014, a special issue of *China Science Bulletin* on information security in China, noted that "very few IT products are designed and manufactured in our country using our own IT, while the majority rely on foreign imports" (Feng 2014: 4162). The cyber security market had become fragmented, with many firms that were lacking in basic competence winning government contracts, a situation that led to a number of unpublicized scandals. The sector was also quite distorted in certain sub-fields. It was dominated by engineering, and to the extent that software got a look-in, the dominant sub-sector was content control, not information security more broadly defined. Cryptography was the single most important non-engineering field of business.

China's business community in this field has been dominated by a clear interest group structure, concentrated around mid-level officials in public security and the state-owned cybersecurity firms, often with links to mass media (because of the heavy concentration on content security and control in China's cybersecurity policies).

3.5   Emergence of Private Firms

The second locus of Chinese expertise is the private sector, including both specialist cybersecurity firms, as well as larger ICT firms that specialize in the consumer market or internet services rather than security. Table 3.1 presents a selection of information pertaining to the security departments of several of the larger commercial enterprises in 2015 (Aqniu 2015).

In 2015, there were around 400 domestic companies in China specializing in cybersecurity (Aqniu 2015). These firms compete against each other against world-leading corporations active in China. In September 2017, a new industry group, the Alliance for Cybersecurity and the Information Industry, was formed under the initiative of the government.

CAICT (2016) reported that in 2015, the operating income of China's listed security enterprises exceeded 10 billion RMB. Its 2017 report gave a more comprehensive picture, including all corporations and not just the listed firms. It assessed that the value of the cybersecurity sector in 2016 was about 34.409 billion RMB, an increase of 21.7% compared with 2015. The projection for 2017 was 45.713 billion RMB (CAICT 2017: 23). At this time, the main contributors in the sector were the SOEs

Table 3.1 Aqniu report on security strengths of selected large firms

Company	Advantages
Ali Security Staff > 2000	information protection, fraud management mechanism, using big data to build powerful real-time risk control and network defence capability
Baidu Security Staff > 1000	world class search technology and state-of-art deep learning technologies, focus on building a trusted and secure internet ecosystem
China Telecom Security Staff > 1000	rich experience, strong network capability, a large number of government and enterprise customer resources, supports hundreds of millions of users
Huawei Security Staff > 2000	sales channels around the world, high annual investment in R&D to ensure the competitiveness of products and future visions
TenCent Security Staff > 1000	ten years of experience from fighting with the "dark industry", analysis of big data systems, quality security services

and the top listed companies, but the private sector of the cybersecurity industry was growing much faster. While there were over 400 cyber security companies in China at this time, the top ten companies comprised more than 85% of the revenue. On the 2017 data, we can calculate that in 2015 the balance between the SOE sector (RMB 28.2 bn) and the private sector (RMB 10 bn), was higher than two thirds/one third (67/33), and that by 2016 this had improved in favour of the private sector by about 12% (59/41). This would represent a substantial shift in favour of the private sector in just one year (assuming the two data sets are compatible).

For the top ten listed firms (essentially the totality of the private sector), the 2017 analysis reported business income the previous year of 14.695 billion RMB, an annual growth rate of growth rate of 34.94% (Venustech, Westone, Bluedon, Utrapower, Bright Oceans Inter-Telecom, Beijing VRV, Surfilter, Lanxum, Meiya Pico, and NSFocus (27). Another 40 privately traded shareholder companies (of which more than 22 had achieved net profit) achieved revenue of 2.788 billion RMB in 2016 (28–29). Thus, the private sector of dedicated cybersecurity companies seems to have represented about half of the cybersecurity industry's business income in 2016. The report did not quantify the activities of the security divisions of companies whose main business is not cybersecurity, such as Huawei, Baidu, Alibaba, China Telecom, and ZTE. Cyber security revenues from these companies are substantial.

The report singled out Wuhan, Chengdu and Shanghai as emerging examples of the clustering effect that created Silicon Valley in the United States, citing in the case of Chengdu a decision announced in April 2016 to invest 13 billion RMB to establish a national cybersecurity base (CAICT 2017: 22) The report also noted that venture capital was an emerging factor and that industry alliances were playing a driving role in the sector (31–33).

A short report in a 2016 yearbook gives some flavour of the evolution (CAICT 2016: 192). In the area of cloud security, the report notes that the National Development and Reform Commission (NDRC) initiated a special project to promote technologies and products, especially security strengthening of the cloud operating system and high performance abnormal flow detection and cleaning. Pilot subprojects were conducted by local governments in Beijing, Shanghai, Shenzhen, and Hangzhou (among others), including the construction of cloud computing centres of different levels, with a special emphasis on security technology. It noted that the 863 program had funded research on data privacy protection, environment monitoring, and security accountability and control for the cloud. Domestic companies, such as AliCloud, Huawei and Qihoo 360, have been cooperating in these pilot projects, notably in devising security solutions in respect of firewalls, intrusion detection, and virus protection. In the area of defence against the increasing threat of APTs in all systems, including mobile and industrial control systems, domestic cybersecurity companies, such as Qihoo 360, HanHai (acquired by Alibaba), Antiy and KnownSec, started research on related defensive technologies concentrating on malware detection, intrusion detection, honey pot, and big data analysis. Of considerable importance, the report noted that "research about threat intelligence analysis and security situational awareness technology has started in China" beginning in 2012. Domestic companies have started R&D on internet security, building intelligent industrial control system security guarantee processes; and strengthening the management of access control management (193). The report concluded that many small companies were competing in this market, presenting a trend of "let a hundred flowers bloom, and a hundred schools of thought contend", a reference to a brief period of pseudo-liberalization under Mao Zedong in 1956.

3.6   Balance Between Foreign and Domestic

In the wake of the Snowden Leaks in 2013, Reuters reported that China's government was preparing to launch a series of investigations of U.S. cyber companies (Reuters 2013). By 2015, Reuters reported that the government dropped some leading foreign technology brands from its list of approved vendors and asked state-owned firms to buy Chinese-made products (Carsten 2015). The Central Government Procurement Centre (CGPC) had almost doubled its approved products list in two years to just under 5000, with the increase reported as almost entirely due to domestic providers and less than half of foreign security-related products remaining on the list (Carsten 2015). Privately-owned Chinese companies joined in, launching what amounted to a 'de-IoE' campaign-a reference to the internet of everything and its globalized character (Cyran 2014). Companies like Alibaba Group followed the lead of SOEs and banks by scrapping hardware and uninstalling software made by American suppliers in favor of domestic brands said to be safe, equally advanced and a lot less expensive (Li et al. 2014). Through this campaign, the market share of foreign firms in the cybersecurity sector declined. Domestic rivals, such as Huawei Technologies and

Table 3.2 Share of market between Chinese and foreign companies

Products		Foreign share (%)
Security hardware	Firewall	27
	Integrated threat management	41
	Intrusion detection system	12
	Intrusion prevention system	34
	Secure content management	20
	Vulnerability management	50
Security software	Identity & access management	35
	Secure content and threat management (including Data Loss prevention)	51

Inspur, won contracts at an accelerating rate. Cisco Systems said the Prism scandal had dented its revenues in China (Mathew 2014a). In October 2013, IBM reported a 22% drop in China sales in Q3 versus Q2 and, according to law suit against IBM, its share value fell by $12 bn as a result (Mathew 2014b). The question here is whether the campaign has been sustained and whether it has led to long-term decline of foreign profit-making in China's cybersecurity sector.

In spite of persistent and impassioned rhetoric about indigenization of the cyber-security sector, China's leaders have not set out a clear set of policies on the acceptable balance between foreign and domestic contributions. In 2014 there was slight improvement in the performance of Chinese companies in some segments of the market, but foreign companies continued to feature prominently, as seen in Table 3.2 (Hui and Tan 2016: 188).

Examination of the change in market share between 2010 and 2014 for eight market segments listed above shows a surge in the performance of Chinese companies, VenusTech, Sangfor and NSFocus, and some notable declines for non-Chinese companies. However the data also shows considerable resilience by leading foreign companies like Symantec and Cisco, while Juniper did not show the same staying power. The comparisons are set out in Table 3.3 which is based on data from Huatai Securities (2015: 45–46), citing a report by the International Data Corporation (IDC).

The table needs to be read with care. By taking 2010 and 2014 as sample years, the analysis misses changes in each of the intervening years, which were more gradual than the five year view suggests. Moreover, annual market share statistics can be volatile while a company's underlying position in the market can be considerably more entrenched. In general however it does seem possible to conclude that there was a shift away from reliance on foreign firms in some segments, most notably the supply of hardware for content management. This latter area, content management, happens to the single highest security priority of the CCP, a view which was increasingly intrusive in political life after President Xi took power in November

Table 3.3 Market share changes for top five in eight market segments 2010–2014

Product	Notable Variations
Firewall	TopSec jumps to 20% (+3.2); VenusTech jumps into top five reaching 9%; H3C grows to 15% (+1); Huawei no significant change; Cisco improves to 8% (+0.7); Juniper drops out of top five (<8%)
Integrated threat management	Cisco drops from lead position at 16% in 2010 out of top five; Venustech races into the lead, with 20% (+11.2); Fortinet and H3C rise into top five; Juniper and Leadsec drop out (<8%)
Intrusion detection system	VenusTech increases its lead at the top to 28% (+8.8); NSFocus boos its second rank position to 22% (+7.4); LinkTrust managed to stay at 4th place with 9% (−2.8)
Intrusion prevention system	NSFocus raced ahead to 23% (+5.5) to keep its first place; TOPSEC came in 3rd at 13% (+5.6) followed by VenusTech which stormed into 4th place at 12% (+6.9); Cisco fell from 3rd to 5th at 6% (−1.5); Juniper dropped out of top five
Secure content management (hardware)	Symantec slumped out of the top five, retreating from 21% in 2010; Kaspersky and Trend Micro also dropped from top five; Chinese firms rushed in with Netentsec at 42%, Sangfor at 23, NSFocus at 8, and H3C in 5th place at 6%; and Cisco eased into 5th spot with 7%
Secure Content threat management (software)	The entire list of top five companies was replaced. Sangfor and Netentsec lost commanding positions to drop out of the top five, surrendering almost 63% of the market between them; with Symantec coming onto the top five and first place with 26%, followed by Trend micro (12%) and Kaspersky (6.5%) at 3rd and 4th; leaving two Chinese companies also new to the top five: Rising at 11% and King Soft at 4.5%; Cisco dropped out of the top five
Identity and access management	Very stable. Chinese companies Jida Zhengyuan (24%) and Koal (16%) remained 1st and 2nd; with IBM (11%) and EMC (9%); and Oracle (7%) coming into the top five to replace Verisign (6.4% in 2010)
Security and vulnerability management	NSFocus stormed into the top five to take 26%; IBM slipped to 2nd place but kept about the same share (22%); Jida Zhengyuan and Koal dropped out of the top five, surrendering 30% of the market share; Venustech came into the top five (13%); HP (9%) and Qualsys (5%) came in at 4th and 5th, displacing ArcSign and Symantec

2012 and before his cyber speech in February 2014. Content control is one of the most politically sensitive sets of cyber operations undertaken by the government. Moreover, the government imposes obligations on many large institutions, such as universities, to undertake their own content control in cyberspace.

The overly hostile approach toward foreign companies ameliorated by April 2016. According to Xi's speech at the national meeting on cybersecurity and informatization in that month (Xi 2016), the government was seeking to project a more moderate attitude towards them. Xi said that "on one hand, core technologies are indispensable for a country" and "must come through self-innovation, self-reliance, and self-improvement, ... from our own research and development. On the other hand, our emphasis on innovation does not mean that we should carry out R&D behind closed doors".

According to a number of sources in China, Xi's speech was the most unambiguous explanation of the country's drive for an indigenous and independent cybersecurity industry but it also signalled a softer attitude toward foreign cyber companies (see for example, Liang 2016). Another source observed that "Though Xi's talk on April 19 was intended for a domestic audience, he responded to outside concerns as well. .... China's Internet sector, he promised, is open to foreign companies" (Lu 2016). In the aftermath of the national meeting, more emphasis was placed on partnership arrangements between foreign and Chinese cyber companies (as opposed to direct unilateral sales by the foreign firms); and the government actively promoted this model to solve information security issues (Liang 2016).

At the same time, the indigenization push remains a powerful and persuasive mobilizer, with much of public policy in the sector organized around it. In his April 2016 speech, Xi pointed out that China lags behind some developed countries in terms of Internet innovation, infrastructure, the sharing of information resources, and the strength of its IT industry. He saw the biggest gap as core Internet technologies. He divided core technology into three categories: basic and generic technologies; asymmetric and "trump-card" technologies; and cutting edge and revolutionary technologies. He also remarked the importance of striking the right balance between openness and autonomy.

At the present stage, China still need to cooperate with foreign corporations in most sub-fields of cyber core technologies. For example, eight US companies—Cisco, IBM, Google, Qualcomm, Intel, Apple, Oracle, and Microsoft, which are often identified by China's state-run media as US government proxies that posed a "terrible security threat," continue to cooperate with Chinese local companies (Tiezzi 2015). These companies all have core technology in some sub-fields.

There have been a number of prominent technologies backed by the government in order to shore up security in cyberspace: two in the area of component development (operating systems and semi-conductors) and two in the area of fundamental scientific breakthroughs (quantum physics and advanced artificial intelligence). The latter is dependent on super computers, a field of technology highly favoured by the leadership, in part for its fit with the classic propaganda line that China has the fastest this or that in the world, or the biggest. China, Japan and the United States have in recent years been vying for the title of the fastest super computer. By way of background, in the mid-1980s the United States loaned to China a Cray supercomputer (then the global leader) but on the condition that it be subject to 24 h video surveillance to prevent any Chinese access to the inner working of the machine. Within 30 years, China has arrived at the top of the tree alongside its American and Japanese counterparts in the speed of its super computers, but this is not the same as the sub-discipline of high performance computing. It is of special note that for practical evolution of this capability and its exploitation in China, several Chinese specialists have observed that "International cooperation is essential" (Xu et al. 2016: 37).

As noted above, in July 2016, President Xi Jinping lamented publicly that China's core technologies remain dominated by foreign firms. This is definitely the case with cybersecurity, even if China is not alone in this. After all, few countries, except perhaps for the United States and Russia, can say that they have an indigenous cybersecurity industry of any scale. For example, the United Kingdom has a bigger cybersecurity industry at $29bn in 2017 (United Kingdom 2017) than China at the estimated $9 bn referred above (51.488 bn RMB), but few countries can match this scale.

On the one hand, foreign companies contribute to the strengthening of Chinese security and indigenous corporate growth in the sector. On the other hand, they have been fighting to maintain market share against the Chinese government's clear protectionist policies and even against the government's efforts to introduce cybersecurity laws similar to those in the United States. China is struggling to make headway in an international environment that favours globalization of the cybersecurity industry, not more indigenization. In spite of the government's promotion of the domestic businesses, it is clear that certain projects can only be delivered by foreign corporations.

Chinese cybersecurity companies are enjoying a rapid growth trajectory. State sponsorship appears to give something of an edge both in the SOE framework, but also in the private sector, where a small group of companies enjoys special security clearance status for obtaining sensitive government contracts. Large firms with business outside the security sector have become major players as well. All of the leading firms are already internationally active, though work to preserve their clearance status to continue to win government contracts. While the state-favoured firms must maintain a 100% Chinese nationality work force, many of them have foreign contacts and relationships that lay the groundwork for further internationalization of the Chinese domestic sector. Domestic firms have been gaining market share in China from foreign firms, but many of the latter group remain active.

Foreign corporations have made an immense contribution to the development of information security in China, and the overall technological development of the country. At the same time, they have been targeted by the government and analysts in China for two things (a) being tools of U.S. intelligence; and (b) legitimate targets for protectionist measures designed to promote an indigenous information security industry. In spite of these very strong headwinds, and a continuing contest over the application the new cybersecurity law which entered force in June 2017, and other security and protectionist regulations, it appears to be the judgement of several leading foreign corporations that they will remain an integral part of the Chinese scene for years to come and that they can even grow parts of their business there.

The inevitably globalized character of the cybersecurity industry and its underlying R&D was demonstrated in December 2017 when Google announced its return to China to open a joint AI research centre (Li 2017). This followed the acrimonious departure from China of the firm when it closed its China search engine because of differences with the Chinese government over censorship policy. Moves by Google to return to China began in earnest in 2016 when the company began hiring in China and was considering a range of business options. The eventual decision by China and Google to agree on a joint laboratory for research on the basic science of AI reflected the new environment for foreign participation: more emphasis on partnership and more emphasis on developing China's S&T base in cyber security. While AI is not exclusively dedicated to cyber security, it is the essential foundation of China's view of it, given the large unequalled scale of the country's security problems in cyberspace.

Scholarly research suggests that foreign firms can mitigate some of the negative pressures, not least through the establishment of R&D centres (Holmes et al. 2016) and through promotion of "reverse innovation" (Zheng et al. 2016). There is a reasonable view that clumsy indigenization policies are sub-optimal in the China/United States relationship and that a "cybersecurity regime [between them] can be complete only if it includes and encourages greater trade and investment between the two countries' technology sectors" (Liao 2016: 44).

3.7   Select Technologies

3.7.1   Operating System

For China, the dominance of Western operating systems (OS) in the global ICT market has been seen as a sign of the country's subordinate position in the sector and something that easily evokes techno-nationalist sentiment. Some see it as implying a natural and pervasive dominance of the English language over Chinese. The dependence on highly vulnerable OS from the West was seen by the government as a wide open door in terms of insecurity. Microsoft Windows has been the dominant operating system in China, and the Chinese government has promoted the development of a domestic OS (Sonnad 2015). The need for this intensified when Microsoft announced in 2013 an end to its support globally for the XP version of Windows. The domestically designed substitute, Kylin OS, was universally regarded to have been both a failure and largely a copy of Microsoft XP, and a new project called NeoKylin was launched in 2015. That effort has been more successful, with one city declaring (perhaps with some exaggeration) that all its workers had begun using it. According to Sonnad (2015) "over 40% of commercial PCs sold by Dell in China" were running NeoKylin. But the Microsoft leadership in Chinese OS choices remains, with the company producing a special version of "Windows 10 China Government Edition" for Chinese government customers (Microsoft Blog 2017) that incorporates special security features specifically requested by the client. In the new regime of partnering, this was in a joint venture with the SoE, CETC.

3.7.2 Quantum Physics

Of the twelve high priority fields for research on the science of information security in the CAS 2011 roadmap that were mentioned above, the Chinese government pays a great deal of attention to quantum computing. According to the CAS, China's quantum communication achievements have "been heralded as the answer to the ever-growing menace of cyber theft" (Xinhuanet 2016). Sources like these regularly report that "devices equipped with these technologies are expected to be ultra-secure as a quantum photon can neither be separated nor duplicated. Scientists say it will be impossible to wiretap, intercept or crack information in quantum communication". China's claims to be the world leader in this field are open to doubt, with some claiming a lead for the United States (Biercuk and Fontaine 2017). These authors, both of whom have strong scholarly credentials to make such an assessment, suggest that "the full promise of quantum technology is unknown, in national security or any other field". So China's claims could usefully be subjected to some further analysis along the following lines.

1. Is China's record in the application, engineering and dissemination of high technology information science strong enough to see it reap the operational cybersecurity benefits claimed for quantum computing? The answer is probably no.
2. Has China spent more or less on the cybersecurity aspects of quantum computing than the US-led alliance? The answer is China has spent much less. In that case, why would we expect China to be in the lead?
3. If China is a leader in quantum communications technology (a sub-field) and it may well be, then how will this manifest itself? Will China want to take the lead in global markets and sell the technology openly, or will it want to try to keep the secrets for itself and apply them exclusively in Chinese military and intelligence applications? China will probably in the end prefer the global, open market outcome.
4. Is China participating actively and expansively in a globalized community of science in quantum computing or trying to keep it all to itself? The answer is probably the former.

3.7.3 Advanced Artificial Intelligence (AI)

As China continued to celebrate its achievements in quantum communication through 2017, it went public with a very sober assessment of its position in the world of advanced AI. In July, the State Council issued new policy guidelines over 28 pages which laid out the ambition to become a world leader in the field while sketching severe limitations on the current state of research and industrial take-up in the country. The grim picture did not stop the document from over-claiming on "China's first-mover advantage in artificial intelligence development". The document identified AI as "the new focus of international competition", noting that it "is thought to be the strategic technology leading the future". The message was that China needed to take the leading position if it was to protect its national security, including internal security, and to exploit revolutionary opportunities for economic gain presented by AI.

While reporting good progress by the country in this field, the document could not have been much more critical. The list of shortcomings was long and represented fundamental gaps, as listed verbatim below:

- lacking major original results in research
- huge gaps in the basic theory, core algorithms, key equipment, high-end chips, major products and systems, components, software and interface
- scientific research institutions and enterprises have not yet formed the ecological circle and production chain with international influence
- lack of systematic research and development pathways
- cutting-edge talent is far from meeting the demand
- the infrastructure, policies and regulations, and standard setting need to be improved.

A recent assessment by the CAICT (2017: 41) was that there had been few achievements in productization in this area of cybersecurity technology in China.

3.8 Conclusion

The political economy of cybersecurity, including its S&T underpinnings, is a subject crying out for advanced research, whether that concerns China or other countries. While China's leaders dream of a high tech future in this field, and its scientists can provide some research underpinnings, the creation (almost from scratch in the late 1990s) of a national cyber industrial complex will have to be a gradual process. Chinese leaders have openly acknowledged this and have publicly presented credible plans for progress. For a number of reasons, the consideration identified by the U.S. NRC and referred to above, that China risks being "designed out of" advanced IT R&D, may appear exaggerated or hyperbolic, but China is somewhat susceptible, as the NRC suggests, to such "self-inflicted wounds" in its national policy settings. Its scientific and industrial communities are increasingly globalized and the only sure direction for China's cyber industrial complex will be, as Xi Jinping says, to intensify that internationalization process-even as he gives strong signals to the contrary in favour of more indigenization.

References

Austin G (2014) Cyber Policy in China. Polity, Cambridge, UK

Aqniu (2015) China cybersecurity companies top 50 hits. 27 Oct 2015. http://www.aqniu.com/industry/11169.html

Austin G, Meng F (2018) Profiles of cyber security companies in China. Discussion brief #5. Australian Centre for Cyber Security. University of New South Wales

Bao X, Xu X, Li C, Yuan Z, Lu C, Pan J (2012) Quantum teleportation between remote atomic-ensemble quantum memories. Cornell University Library. submitted 13 Nov 2012. http://arxiv.org/abs/1211.2892

Biercuk M, Fontaine R (2017) The leap into quantum technology: a primer for national security professionals. In: War on the rocks. 17 Nov 2017. https://warontherocks.com/2017/11/leap-quantum-technology-primer-national-security-professionals/

CAICT (2016) Research on the competitiveness of global cyber and information security businesses. The Security Research Institute of China Academy of Telecommunication Research. In Chinese. In: Hui Z, Tan Q (eds) Annual report on development of cyberspace security. Social Sciences Academic Press, Beijing, pp. 176–200

CAICT (2017) White paper on cybersecurity industry (2017), In Chinese. China Acad Inf Commun Technol. Sept 2017. http://www.caict.ac.cn/kxyj/qwfb/bps/201709/P020170919308653198647.pdf

Carsten P (2015) China drops leading tech brands for certain state purchases. Reuters. 27 Feb 2015. http://www.reuters.com/article/us-china-tech-exclusive/china-drops-leading-tech-brands-for-certain-state-purchases-idUSKBN0LV08720150227

China.com (2015) Vanguard of national information security. In Chinese. 22 Jan 2015. http://tech.china.com/news/it/11146618/20150122/19236241.html

Cyran R (2014) China's 'De-IOE' campaign takes a bite out of tech. Reuters. 16 July 2014. http://www.reuters.com/article/idUS122258395120140716

DRC and World Bank (2013) China 2030: building a modern, harmonious, and creative society World Bank Publications Development Research Centre of the State Council

Feng D (2014) Preface. China Sci Bull 59(32):4161–4162. http://www.scichina.com:8080/kxtbe/EN/abstract/abstract509393.shtml

Holmes RM, Li H, Hitt MA, DeGhetto K, and Sutton T (2016) The Effects of Location and MNC Attributes on MNCs' Establishment of Foreign R&D Centers: Evidence from China. In: Long Range Planning. 49(5):594–613

Huatai Securities (2015) Research report on the cyber security Industry. In Chinese. Tencent. 16 Sept 2015. http://new.qq.com/cmsn/20150917031679

Hui Z, Tan Q (2016) Cyberspace security in the era of data economy: global and Chinese contexts. In: Hui Z, Tan Q (eds) Annual report on development of cyberspace security in China. Social Sciences Academy Press. Blue Book, Beijing. (In Chinese). 1–16

iiMedia (2017) 2016 China internet security research report. In Chinese. http://www.iimedia.cn/148999407414508836.pdf

Li F (2017) Opening the Google AI China center. Google blog. 13 Dec 2017. https://www.blog.google/topics/google-asia/google-ai-china-center/

Li G (ed) (2011) Information science and technology in China: a roadmap to 2050. Chinese Academy of Social Sciences, Science Press, Springer, Beijing

Li J (2015) Network information security challenges and relevant strategic thinking as highlighted by "PRISM". In: Huang Z Sun X Luo J, Wang J (eds) Cloud computing and security. Lecture notes in computer science, vol 9483, pp 147–156. Springer, Berlin

Li X, Qin M, Zhang Y, Nan H, Qu Y, Zheng P (2014) China pulling the plug on IBM, Oracle, others. MarketWatch. 26 June 2014. http://www.marketwatch.com/story/china-pulling-the-plug-on-ibm-oracle-others-2014-06-26

Liang C (2016) Network information security shift 'to foreign capital'. Caijing 12 Aug 2016. In Chinese, Foreign investment in China, p 8. http://magazine.caijing.com.cn/20160812/4161944.shtml

Liao R (2016) Dysfunction, Incentives, and Trade: Rehabilitating US-China Cyber Relations. In: Georgetown Journal of International Affairs. 17(3):38–46

Lindsay JR (2017) Restrained by design: the political economy of cybersecurity. Digital Policy, Regul Gov 19(6):493–514. https://doi.org/10.1108/DPRG-05-2017-0023

Lu C (2016) China's emerging cyberspace strategy. The Diplomat. 24 May 2016. https://thediplomat.com/2016/05/chinas-emerging-cyberspace-strategy/

Lu Y (2010) Science & technology in China: a roadmap to 2050. Strategic General Report of the Chinese Academy of Sciences, Springer

Mandiant (2013) APT1: exposing one of China's cyber espionage units. https://www.fireeye.com/content/dam/fireeye-www/services/pdfs/mandiant-apt1-report.pdf

Mathew J (2014a) Cisco's China equipment sales dented by NSA spying scandal. International Business Times, California. 1 July 2014. http://www.ibtimes.co.uk/cisco-china-sales-nsa-spying-scandal-522159

Mathew J (2014b) IBM sued over China sales decline due to NSA 'connection'. Reuters. 1 July 2014. http://www.ibtimes.co.uk/edward-snowden-nsa-scandal-ibm-lawsuit-530150

Marks J (2015) U.N. body agrees to U.S. norms in cyberspace. Politico. 9 July 2015. http://www.politico.com/story/2015/07/un body-agrees-to-us-norms-in-cyberspace-119900

Microsoft Blog (2017) Announcing windows 10 China government edition and the new surface pro. Microsoft Blog. 23 May 2017. https://blogs.microsoft.com/blog/2017/05/23/announcing-windows-10-china-government-edition-new-surface-pro/

Nordum A (2017) China demonstrates quantum encryption by hosting a video call. IEEE Spectrum, 3 Oct 2017. https://spectrum.ieee.org/tech-talk/telecom/security/china-successfully-demonstrates-quantum-encryption-by-hosting-a-video-call

NRC (2010) S&T strategies of six countries: implications for the United States. National Research Council. National Academies Press. https://www.nap.edu/read/12920/chapter/6

Pasick A (2015) It's official—China is blacklisting Apple, Cisco, and other US tech companies. Quartz. https://qz.com/351256/its-official-china-is-blacklisting-apple-Cisco-and-other-us-tech-companies/

Patton D (2015) U.S. tech firm Cisco to invest $10 billion in China expansion. Reuters. 18 June 2015. https://www.reuters.com/article/us-china-Cisco/u-s-tech-firm-Cisco-to-invest-10-billion-in-china-expansion-idUSKBN0OX1ZP20150617

Reuters (2013) China to probe IBM, Oracle, EMC for security concerns—paper. 16 Aug 2013. http://www.reuters.com/article/china-ioe/china-to-probe-ibm-oracle-emc-for-security-concerns-paper-idUSL4N0GH06O20130816

Russian Federation (2015) Order of the Russian Government on signing the agreement between the government of the Russian Federation and the Government of the People's Republic of China on cooperation in the field of ensuring international information security. In Russian. 30 Apr 2015. http://government.ru/media/files/5AMAccs7mSlXgbff1Ua785WwMWcABDJw.pdf

SCIO (2015) China's military strategy. White paper. State Council Information Office. http://english.gov.cn/archive/white_paper/2015/05/27/content_281475115610833.htm

Sinolink Securities (2017) Bright star stock analysis: significant advantages of the industry boom company, year-round business trends to improve. In Chinese. http://www.microbell.com/docdetail_2117399.html

Sonnad N (2015) A first look at the Chinese operating system the government wants to replace Windows. Quartz. https://qz.com/505383/a-first-look-at-the-chinese-operating-system-the-government-wants-to-replace-windows/

Symantec (2015) Symantec 2015 annual report. Available at: http://s1.q4cdn.com/585930769/files/doc_financials/2015Report/SYMC-2015-Annual-Report-Bookmarked-FINAL.pdf

Tiezzi S (2015) New report highlights China's cybersecurity nightmare. The Diplomat. http://thediplomat.com/2015/02/new-report-highlights-chinas-cybersecurity-nightmare/

United Kingdom (2017) Matt Hancock's cyber security speech at the Institute of Directors conference. 27 Mar 2017. https://www.gov.uk/government/speeches/matt-hancocks-cyber-security-speech-at-the-institute-of-directors-conference

WEF (2016) Global information technology report 2016. World Economic Forum. http://www3.weforum.org/docs/GITR2016/WEF_GITR_Full_Report.pdf

WEF (2011) Global information technology report 2010–2011. World Economic Forum. http://
    resource-cms.springer.com/springer-cms/rest/v1/content/36522/data/v3

Wu Q (2016) Grasp the strategic opportunity to promote autonomy and control, to play the historical
    responsibility of network security. In Chinese. http://www.capco.org.cn/content/29706.shtml

Xi J (2016) Speech at the work conference for cybersecurity and informatization, Posted on 19
    Apr 2016, updated on 26 Apr 2016. https://chinacopyrightandmedia.wordpress.com/2016/04/19/
    speech-at-the-work-conference-for-cybersecurity-and-informatization/

Xinhua (2016a) President Xi says China faces major science, technology 'bottleneck'. 1 June 2016.
    http://news.xinhuanet.com/english/2016-06/01/c_135402671.htm

Xinhua (2016b) Xi sets targets for China's science, technology progress. 30 May 2016. http://news.
    xinhuanet.com/english/2016-05/30/c_135399655.htm

Xinhua (2016c) President Xi's speech on science, technology published. 2 June 2016. http://usa.
    chinadaily.com.cn/china/2016-06/02/content_25595722.htm

Xinhuanet (2016) Digital stars in spotlight at World Internet Conference. Wuzhen. Available: http:
    //news.xinhuanet.com/english/2016-11/16/c_135835057.htm

Xu Z et al (2016) High-performance computing environment: a review of twenty years of experi-
    ments in China. Natl Sci Rev 3(1):36–48, 1 Mar 2016. https://doi.org/10.1093/nsr/nww001

Yu H (2017) Reading the 13th five-year plan: reflections on China's ICT policy. Int J Commun
    11(2017):1755–1774

Zhang S (2016) China sets goals of informatization. CRI News. 28/07/2016. http://english.cri.cn/
    12394/2016/07/28/3821s935816.htm

Zheng Z, Chen J, and Zheng G (2016) Typical modes and characteristics of reverse innovation: a
    multiple cases study on five enterprises operated in China. In: International Journal of Technology
    Management. 72(1–3):43–60

Chapter 4
Corporate Cybersecurity

Abstract This chapter looks at trends in how well prepared Chinese corporations are to defend themselves in cyberspace at home. It looks in broad terms at the question of security culture in enterprises, and at the special case of the financial services sector, especially banks. This sector was the main initial focus of government policy for informatization at the turn of the century (Austin in Cyber policy in China. Polity, Cambridge, 2014: 94). Other sectors of note in this chapter include airlines, the electricity grid and universities, though these are discussed only in brief.

China's official propaganda creates the impression that the only corporations inside the country that need to be protected in cyberspace are Chinese. The mantra has been that the government must enhance national cybersecurity capability to protect sovereign interests and provide "Chinese cybersecurity defences" for Chinese firms. This is only part of the story of corporate cybersecurity in the country. The other more complex part of the story is that the fabric of corporate life in China is highly internationalized. There are many elements to this second part. Chinese entities, like the Bank of China and the country's air traffic control network, have to interface every day with foreign computerized systems which are protected by non-Chinese cybersecurity systems. The same is true of Chinese e-commerce firms and universities, among many others. At the same time, the ICT manufacturing sector in China is comprised in large part of foreign-invested enterprises and in some cases wholly owned foreign enterprises. Leading Chinese enterprises, such as the Bank of China, have offices and highly capable cyber security staff outside the country. This bank is one of the most sensitive financial institutions in the world given its management of the largest single reserves in the world, the underwriter of the locomotive economy of the world, and its role as one of the four largest economies in the world (alongside the United States, the European Union and Japan).

As a result of these considerations, it is not possible to see the cybersecurity of corporations in China, and their business interests, as an exclusively Chinese affair. The client lists of leading Chinese and foreign cybersecurity companies include both Chinese and foreign corporations, and in both cases, regardless of nationality of the entities, they might be operating outside of China. It may not be possible any longer, if it ever was, to accurately document the scale of the cybersecurity business conducted from inside China that is directed to Chinese corporations; or to Chinese corporations

G. Austin, *Cybersecurity in China*, SpringerBriefs in Cybersecurity,
https://doi.org/10.1007/978-3-319-68436-9_4

operating outside China. For example, the decision announced in September 2017 for China Telecom Global Limited to partner with an Indian firm, Versa Networks, as its primary vendor for software defined wide area networks for 15 cloud hubs around the world (Ziser 2017) seems to defy easy analysis according to a techno-nationalist lens, not least because it is part-invested by Verizon ventures, headquartered in the United States.

To complicate this type of assessment, there are few reliable estimates of the cost of cyber insecurity to Chinese corporations. Following the lead of academic studies outside China (for example Anderson et al. 2013), these could come in several forms:

- The actual balance-sheet value of losses incurred directly as a result of cyber-enabled traditional crimes, such as theft of funds or fraud
- The actual balance-sheet value of losses incurred directly as a result of cyber-dependent crime (attacks against networks, systems and data)
- The estimated balance sheet losses arising from damage to reputation or brand value incurred as a result of the above
- The estimated balance sheet losses arising from theft of intellectual property (the implied loss of past or future earnings)
- The share of enterprise revenue that has been spent to prevent cyber crime and maintain information security.

As Anderson et al. (2013: 266) point out, there may be other ways of understanding the costs, such as net criminal revenue from cyber crime or some imputed cost to the society as a whole. For the purposes of this chapter on corporate responses to cybersecurity challenges in China, we can focus on the four bullet points above. Each corporation will know of its balance sheet losses but the government appears to be capable of tracking only such losses of the most serious kind. It has no formal mechanisms in place to track all instances of cyber crime. It does not publish comprehensive data on the costs to corporations of the cybersecurity mission.

China does have a wide range of data on what McGuire and Dowling (2013: 6) call negative experiences in cyberspace (i.e. cyber hacks or virus infections), but many of these (an unknowable number) would not involve recordable losses or costs or be registrable as crimes worthy of investigation. Warner and Sloan (2016: 3) concluded that "there is no alternative to finding some way to provide the currently unavailable information", and that mandatory data breach reporting is not the answer.

Wolf and Lehr (2017: 1) conclude that many of the data gaps will endure, but that useful policy decisions can be made by corporations (and governments) "so long as the government can encourage a flow of good third-party research". They observe correctly that much of the data needed will be in the hands of information security providers (and insurers). Part of these judgements holds for China, but we can note that there is very little research in China on these topics and no domestic cyber insurance effort to speak of. China's lack of research is a shared global problem since the study of cyber crime "is still in its infancy" (Armin et al. 2015: 709). A particularly important point to note for the purposes of this chapter is that the dollar costs of cyber insecurity are highest for business (TNO 2012), a proposition that earns a mention in a short study sub-titled "No Country Is Cyber ready" (Hathaway 2013: 2).

4.1   Corporate Response to the Cybersecurity Challenge

As disscused in Chap. 7, the overall assessment by the Chinese government is that the cybersecurity situation in the country is stable, but that there are some worrying trends, including rampant cyber fraud. One trend the government has not called out as fully as it might have is the dismal record of Chinese corporate entities (such as businesses, universities, lawyers, doctors, trade unions, non-government groups) in efforts to provide for their own security. There are clear exceptions, as mentioned for example in the banking sector.

The Chinese government has been tracking corporate responses for more than 20 years in surveys conducted by CNNIC. According to the 39th statistical report in 2017, CNNIC was able to report that by the end of 2016, most companies (over 90%) have taken basic cybersecurity measures (CNNIC 2017: 20). The CNNIC survey data for that year indicates that only 4% of enterprises pay for anti-virus software and firewall protection, and that only half of all companies actually use such technology (46% use free products). By December 2016, just over one fifth of companies (22.3%) used integrated hardware software protection, and another one tenth of companies (9.5%) had deployed hardware protection systems independently of their software systems, though uptake on the first count was marginally higher than in 2015 (a 6% rise). On average, more than 40% of medium and large sized companies (those with more than 100 personnel) did not have any dedicated network management staff (20–21). For the smallest firms (one to seven people), the proportion of firms without such dedicated staff reached 70%. In March 2017, Gartner reported that "Although security, safety and risk rank among the top 10 business priorities of global enterprises, they do not feature in the top 10 priorities of Chinese enterprises" (Gartner 2017a). This report also mentions that only 8% of Chinese firms surveyed indicate that "security is a technology priority", and that this is about half of the global average.

Qihoo 360 reported very high rates of attack in 2016, most of which target private enterprises of one kind or another or government agencies (Qihoo 360 2017a, b: 15). In its 2016 threat report, Qihoo also estimated that most entities (just over 50%) maintain highly vulnerable websites and that these were being subjected to around 5 million attempted attacks per day (10). Qihoo assessed that most of these attacks appeared to have come from inside China, primarily Jiangsu, Beijing and Henan. In terms of scanning attempts (looking to find vulnerabilities), Qihoo reported that the main intrusions came from the United States, Russia and Brazil. Corporate users in China were receiving an average of more than 20 million spam messages a day, accounting for around 70% of the total number of e-mails sent to corporate users (120). Spam messages provide a cover for attacks seeking access to user or administrator data or for initiating malware attacks.

In 2016, some 36 APTs were found to have been monitoring around 200 targets in China. Universities accounted for the highest proportion of targets (40%), followed by enterprises (25%), and government agencies (18%) (Qihoo 360 2017a, b: 136). Qihoo assessed that enterprises still have a lot of blind spots in network security construction when it comes to defence against APTs and that domestic dynamic security providers "are still seriously out of position" with respect to that threat (144). Qihoo observed that the "data driven, coordinated and linked, defence system will become the main method of APT detection and defence in the future".

Qihoo found that for Chinese enterprises involved in around 500 security incidents handled by it in 2016, that only 5% of the incidents had been independently identified by the victim companies (144). In another 27% of cases, the attacks were identified only after significant signs of invasion or economic losses had occurred. The other 68% of attacks had not been discovered by the company until the 360 CERT informed them.

Will corporations follow the indigenization market signals of the Chinese government? According to a Gartner estimate, they will: "more than 80% of large businesses in China will be deploying network security equipment from local vendors by 2021" (Gartner 2017b). While that may well happen, the Gartner estimate does not make clear what proportion of the total cybersecurity system in each company that indigenous component will represent. In many cases, it may well be the smaller part. One alarming trend called out by PWC is that in spite of increasing threats and a 969% increase in incidents for a set of companies surveyed in China in 2016, their total cybersecurity budgets in 2016 "had dropped 7.6% from the previous year" (Soo 2016).

4.2 Organizational Culture and Employees

Corporate practices with respect to the place of organizational culture and their employees' security behaviour are relatively underdeveloped in China compared with many other G20 countries. This reflects two circumstances: that management studies in China is a late-comer discipline in the universities; and that it is an even newer field of research when it comes to management of information security practices (Zeng et al. 2013: 684). It was only in June 2017 that the government set up the Industry Information Security Alliance which, according to CAICT (2017: 33) "accepts the guidance of the Ministry of Industry and Information Technology" and in 2017 comprised "149 member companies, including Shenhua Group, ChinaCar Group, Aviation Industry, China National Armed Forces, China Electronics Information Industry Group and other industrial leaders".

A useful recent case study in China of "information security culture" in a large manufacturing company presented some preliminary results (Tang and Zhang 2016). The work, funded by the Ministry of Education, used the largest garment manufacturer in China, supplying major international brands such as Nike, Adidas, Ralph Lauren, and Tommy Hilfiger, for a case study. The results suggested that a process orientation (as opposed to a results orientation) was believed to promote better compliance with cybersecurity management regimes (184). There was also more faith for cyber security compliance in a culture that looked to the mindset of workers rather than to duties of the job; one that had open communication systems (not restricted)

(84–85); one that was more parochial than professional (185); one that favoured tight control over loose control; and one that was more normative than pragmatic (customer oriented) (185). The results are case-specific, but suggest that a stronger "supervisory" element in work culture as a whole is preferred by the people interviewed to an approach that might be more in line with general business practices in Western countries.

In another study, this one designing a theoretical approach for use in Chinese firms that have received state certification for their cybersecurity, three researchers suggest that businesses can develop employment contracts that use incentives to maximise employee compliance with information security regimes (Lin et al. 2016). They suggest that only coercive pressure can bring employees to internalise the values of these regimes but that if coercive measures are used, there is not as much positive impact as use of incentives.

Whatever the strengths of such research in China on corporate culture and cybersecurity, there is little evidence of its widespread take-up.

4.3 Financial Services Sector

The leading banks in China are well prepared for cyber threats but the rest of the financial services sector in China remains one of its most vulnerable for criminal attack because of system-wide vulnerabilities and the opportunity for higher criminal gain. Chinese authorities have identified that almost three quarters of cyber attacks on China in 2011 used websites posing as Chinese banks (Tang et al. 2016: 179). In 2015, the Guiding Opinions on Strengthening Financial Consumer Protection (Guo Ban Fa No. 81, 2015) include the "right to information security" as one of eight consumer rights (CBRC 2016: 133).

The Bank of China (BoC) has taken a lead role in developing cybersecurity practices. In 2006, in its regulatory role for the banking sector, BoC issued Guidance on Further Strengthening the Information Security of Financial Institutions. In terms of its own security, according to the bank, it deploys a multi-level network security protection system with a diversified target set. The system is based on a "three centres" architecture linked by high-speed optical fibre: two are in separate places in Beijing (the main centre and a back up centre) and the third (for disaster resilience) is in Shanghai. The bank implements a multi-level security strategy based on access policy, the isolation of networks by function, and monitoring. According to the bank, the 2016 Cybersecurity Law mandates this approach for all financial institutions, even though it was not defined further in the law. The bank also reports that it has a wide range of other world standard security processes in place, including vulnerability scanning, inspection of operating systems, timely discovery of configuration irregularities, and timely repair of faults (BoC 2016: 4–5).

The BoC has also imposed other general obligations on enterprises, including preservation of internet logs for a minimum of 6 months, backup and encryption of key data, with financial services firms needing to appoint dedicated security management bodies and security-vetted personnel (Lou et al. 2016).

The Bank operates globally and trillions of virtual dollars pass through its systems every week. Its recruitment practices for information security specialists outside China are largely non-discriminatory as to nationality, and job specifications appear to conform to global best practice, with compliance and risk management being two of the primary non-technical responsibilities in such roles. Compliance in the case of the bank is double-sided and involves conformity with both local laws and internal bank procedures. Former employees of the Bank of China in information security are sprinkled throughout global infosec companies or in related roles in non-infosec companies. In late 2016, the Bank advertised for six specialists in security and cryptocurrencies with PhDs or Master's degrees, to take forward its research in that area that began in 2014 (Young 2016).

The Bank is serviced by several Chinese and international IT or infosec providers, including IBM. In 2014, as mentioned in the previous chapter, reports circulated that Chinese banks had been instructed to get rid of foreign equipment on security grounds, but the actuality appears to have been that they were asked to report on the security impacts of using foreign equipment and services (Reuters 2014). Reuters reported that sources in the four major state-owned banks rejected the reports of an order to ban IBM servers.

In recent years there have been reports of successful but minor attacks on the BoC. In 2013, the Bank suffered a DDOS attack of unknown impact in reaction to its change in policy on Bitcoin exchanges in the country (Charlton 2013). In 2015, its Hong Kong branch was subjected to website defacements and DDOS attack but neither affected its financial operations or client data, according to the Bank (CNN 2015). Of some note, the SAR's Monetary Authority required it to submit a report on the attacks, a public obligation not shared by banks in the rest of China. In May 2017, when businesses around the world were affected by the ransomware Wannacry, the Bank took its ATMs off line (Reuters 2017) but it was not clear whether this was a precaution or because they were infected (as were the automatic payment machines in one of China's petrol distributors).

The China Banking Regulatory Commission (CBRC) has been taking its responsibilities of oversight of information security seriously, and its annual reports in recent years have mentioned a range of activities. Its 2012 Annual Report gives some feel for how underdeveloped the situation in some banks may have been. It reported that the Shandong provincial office of CBRC "allocated personnel to track national notifications on information security vulnerability, and advised locally incorporated banks to establish e-banking risk assessment mechanisms" (CBRC 2013: 72). It also reported that the same office implemented "24-h real-time monitoring over phishing sites and established the mechanism for emergency communication" with CNCERT/CC "and domestic information security companies". In 2014, CBRC referenced an important instruction issued in September, Guiding Opinions on Strengthening the Banking Cybersecurity and Information Technology Construction through the Application of Secure and Controllable Information Technologies [Yin Jian Fa, No. 39, 2014] (CBRC 2015: 115). As a result, the Commission launched a "strategic alliance" within the sector for secure and controllable information technology and set up central technology laboratories to support it and promote innovation. In December 2014, the CBRC released formal Guidelines for Application of Secure and Controllable Information Technology in the Banking Industry.

In its 2015 Annual Report, CBRC noted its continuing work in this area:

- risk consultations for national-level banking institutions
- security risk penetration testing based on the scale and classification of banking institutions
- self-evaluation by banks of security of mobile-banking
- quality control of information security internal procedures
- promotion of information sharing and rapid response (CBRC 2016: 119).

In May 2015, the National Internet Finance Association of China (NIFA) announced it was setting up a Special Committee for Internet Finance Cyber and Information Security.

According to Qihoo 360 (2017a), the number of vulnerabilities and the number of high-risk vulnerabilities in the financial sector in 2016 were among the highest in all industries. In the first 11 months of the year, the number of vulnerabilities in financial websites (over 1700) and the number of high-risk vulnerabilities (about 700) was higher for this sector than its closest 'competitors' (education, road transport, and health care). Websites of various sub-sectors of the financial services sector, such as insurance, had particularly bad situations. More than 260 vulnerabilities were reported in the insurance sub-sector, the banking sub-sector reported more than 130 vulnerabilities, and the securities sub-sector report 70 vulnerabilities. One of the vulnerabilities revealed in the website of an insurance association of China in April 2016 had the potential to leak personal data on 800 million policies involving hundreds of millions of clients. Emerging financial businesses were found to have many problems in website security. Moreover, a number of third-party payment companies were also found to have a number of vulnerabilities. APT attacks in the financial services sector in 2016 further complicated the threat to commercial banking networks.

A 2016 study of financial institutions in the Greater China area (including Hong Kong and Taiwan) analysed organizational responses and the online security performance (OSP) of staff (Li 2015). It is useful for its findings, but it is also especially useful in indicating the state of play at the time and some of the character of the changing environment. Li examines the compliance of the institutions with common regulatory guidance in the three jurisdictions on the adoption of two-factor authentication (2FA) (e.g. a standing password and a one-time password generated remotely and communicated to the user's phone or email). The study notes that in spite of such advice being in place for several years, there were no public studies of compliance before this one (Li 2015: 20). The study showed that 65.2% had applied 2FA. The sub-sector breakdown was very telling: 98% of banks, 0% of insurance firms, and 43% of securities trading firms (Li 2015: 23). No biometric authentication was used by any of the participating respondents. The study also found that while a large majority of institutions mentioned information security in the annual report, only

20% had any notable additional discussion of it, for example in terms of business continuity, and that none discussed expenditure on information security or security breaches and intrusions (Li 2015: 24). Li goes beyond the 2FA question and develops a typology for OSP that comprises 16 additional elements, all related to authorization and authentication.

It should be noted that the financial services sector is a particular priority for China's quantum computing research. When Pan Jiwei set up the "world's first large-scale metropolitan area quantum communication network" in Hefei in 2009, it linked a number of government agencies, financial institutions and research institutions (CAS 2014: 63). The team also developed a "quantum validation network for financial information security" in collaboration with Xinhua News Agency, "applying quantum communication network technologies to the secure transmission of financial information for the first time in the world".

4.4 Airlines

China has been more alert than most countries to the threat of cyber attack on civil aviation, having hosted the annual conference of the International Civil Aviation Organization (ICAO) in 2010 when it voted to adopt two new international conventions on terrorist action against aircraft, that included provisions relating to "technological attack" (meaning cyber attacks among other things) on aircraft or air traffic control systems. (Austin et al. 2014: 17–18).[1] This leading position of China was confirmed in February 2017, when the country's Legislative Affairs Office (LAO) released the draft Interim Provisions on Administration of Network and Information Security in Civil Aviation (USITO 2017). According to USITO, this was the "first sector specific network information security policy" released by the LAO "in accordance with the Cybersecurity Law (CSL)". USITO reported that the accompanying explanatory memorandum from the LAO noted the dependence of civil aviation on the country's CII.

In October 2016, China had participated in the ICAO Assembly which noted the need for action to address cyber risks to the industry in the light of the 2010 treaty signings (Resolution A39-19) (ICAO 2016). In April 2017, China participated in the inaugural ICAO summit on cybersecurity, including industry representatives, that was called to discuss implementation of the 2016 resolution and which issued an "industry" Declaration on Cybersecurity in Civil Aviation (ICAO 2017). The document affirmed the multi-disciplinary character of cybersecurity and the centrality of international collaboration to address the challenges. One of these called out by the President of the ICAO Council, Dr Olumuyiwa Benard Aliu, was that "traditional first responders in aviation, pilots and air traffic controllers, must have the training and capabilities they need to recognise and respond to cyber-attacks and effectively

---

[1] As of 1 September 2017, China had not ratified these treaties and they had not entered into force since the number of ratifications had not yet reached 22.

intervene in case of system failures". This is a useful reminder of the sort of training and education systems that China (like all countries) will need to put in place to achieve even basic safeguards for the contingency of cyber attack on aircraft in flight or on the systems supporting them. Some Chinese airlines appear to have moved robustly before these meetings to establish cyber-risk management systems but the public domain evidence is very sparse.

In 2016, according to its publicity, China Eastern Airlines "constantly promoted informatization infrastructure construction and established infrastructure assurance and globalization service systems to ensure a stable information system" and it "built a three-dimensional information security system" (network protection, access control, and monitoring) (China Eastern 2017: 23). In October, 2016, the information work team of China Eastern won the first prize in China's 1st Civil Aviation Internet Security Protection Skill Competition. As an airline in possession of massive passenger information, the Company says it has "complied with laws and regulations and occupational norms, valued passenger information protection, and formulated [appropriate] regulations".

According to Xinhua (2016), Capital Airlines signed a strategic cooperation agreement with Qihoo 360 enterprise security group to launch comprehensive collaboration on information security. The news report noted that "in the face of the serious situation of aviation network security, this is the latest positive action taken by Chinese civil aviation enterprises and cooperation with network security companies". The agreement would address five aspects: application system evaluation, system testing, source code auditing, safety notices, and emergency response. The news item reported that official studies assessed the level of information security sector as less advanced than that in "insurance, banking, logistics and other industries". It noted that in 2016 the Civil Aviation Administration of China (CAAC) has issued a work notice on the safety inspection of network systems and called for the airline companies to do better.

In June 2017, China Southern introduced facial recognition technologies instead of boarding passes for passengers to access flights (Zheng 2017). The technology matches the image taken on arrival with the photo in the passenger's official ID document and other headshots that can be uploaded in advance on the airline's mobile app. The move has far-reaching implications for real-time surveillance and tracking of Chinese citizens by internal security agencies, and also raises concerns about the security of the imaging data, but it also reduces the (small) risk of fraudulent use of airline passenger information systems.

4.5 Electricity Grid

As early as May 2002, the central government announced mandatory provisions on the information security of grid and power plant systems and networks (Chen et al. 2013: 544). By that time, Chinese researchers had become seriously engaged with information security aspects of electric power management. Through the following decade, as interest in "smart grid" management gathered pace, so interest in securing the new systems also gathered pace. In 2011, State Grid Corporation of China (State Grid) had put in place its first large scale information disaster recovery plan (SGCC 2011). That year, it also established the State Grid Research Institute, a small centre with some emphasis on information security, headquartered in Beijing, with branches later opening in Berlin (2014) and Santa Clara (California).

The challenge is massive. On the one hand, as a state-owned enterprise, State Grid operates the power distribution for 26 provinces (or province-level municipalities), or 88% of Chinese territory. On the other hand, the organization is federated in practice (with many subsidiaries) originating in a variety of provincial level arrangements, since State Grid was set up only in 2002. The company must not only prevent and remediate cyber attacks, but it must ensure the stable and secure operation of the power grid itself and its information systems. Its purview includes dispatch automation, production management, marketing management, power supply services, and e-commerce. It is especially challenged by the continuous roll-out of new technologies (cloud, Internet of things), but its reliance for smart grid services depends on big data analytics that present their own information security dilemmas.

Thus the biggest utility in the world has been saddled with a dual task: computerize a sprawling and massive network serviced by people in remote areas of China which are known to be less digitally literate (cities like Lhasa, Urumqi and Haikou), at the same time as enforcing effective security information standards. As an indicator of the scale, in an initial webinar on information security sometime in 2016/17, the number of participants reached 15,000 (State Grid 2017).

One indicator of the departure point and timing was a study in 2012 reporting that "most of the electric power information systems use plaintext transmission", a situation which "leads to electric power data more easily modified, monitored, forged, and deleted" (Ran et al. 2012: 144). By 2015, information security plans had been announced for State Grid and regulatory agencies had laid some information security regimes for smart grid development, including standards, information partition and hierarchical authorities (Brunekreeft et al. 2015: 182).

There have been few reports of breaches in the system. In 2016, there was a report of data leakage in State Grid's official App that potentially exposed data on more than 10 million customers. While the company denied the leak, CCTV (2016) reported a knowledgeable source who explained that the problem occurred because of password sharing in the rush to get customers signed up for e-commerce applications ("handheld power") and because different regions had different rules. The information security transformation was proving to be a work-in-progress. Writing in 2016, one researcher described the security challenges facing State Grid as unprecedented, and proposed attention to six key technologies: encrypted trusted access controls (including revocation of access); secure indexing to "achieve accurate ciphertext retrieval and processing"; "user-oriented method of separately verifying the data retrievability proof"; adoption of an American proposal for client data privacy in analysing use; automated systems for "file access control, source tracking, filtering and scanning, … and [to] detect and fix security breach issues"; "unified identity authentication management of users who access shared resources", especially where access crosses many domains; and adoption of a trusted cloud computing platform (Shen et al. 2016: 762–764). Once might presume from this analysis that such highly mature approaches were not in place in State Grid across its entire enterprise.

In 2017, citing the anniversary of the prompt delivered to the nation by President Xi in 2016 to do better in cyber security, State Grid acted (State Grid 2017). The Chair of the Board (also its Party Secretary, and leader of its network security and informatization leading group), Shu Yingbiao, announced that the company had been very active in the year since the Xi speech. Shu was appointed to the leadership role only in 2016, after a career in the company (and its predecessors). The list of actions announced by State Grid was long and contained many breakthroughs that will take years to bear fruit:

- sorted out the "Ten Major Issues and Four Challenges" faced by the company's cybersecurity work
- published a "Network and Information Security Report" (presumably the first)
- initiated major projects to strengthen the security management and oversight system and technology upgrade (involving a total of 31 key tasks)
- carried out a 3-month large-scale safety inspection of power monitoring and control systems completed 20 rectification and management tasks
- changed its management structure and compacted its responsibilities at all levels
- each "unit" in the company set up a leading group for network security and informatization (mirroring the Central Leading Group chaired by Xi)
- set up more than 1500 cyber security management offices and full-time positions at all levels
- set up a network security management post in core business units, such as transportation, inspection, and marketing
- introduced annual performance appraisal standards of the person in charge as the starting point
- increased network and information security assessment
- improved the analysis of information security incidents
- optimized network isolation and terminal access policies.

State Grid also committed to "promote the research and application of new technologies such as trusted computing and quantum communications, and complete the research and development of new encryption chips and a new generation of special protection equipment".

A news report in August 2017 revealed a new intensity for information security in the electric grid (Xinhua 2017). It described 'blue team' building in the Chongqing Electric Power company to strengthen information security. The role of this team is to monitor attack behavior, construct defenses, and accelerate the implementation of security measures (including through training). The report talks of a "rapid rise" in the capabilities of the Blue Team. The article held up the Chongqing activity as a model for the rest of the country.

4.6   Universities

One interesting case study for cybersecurity in any country is the university sector, one of the sectors in advanced countries most attacked by states in cyber espionage. They are a prime intelligence target because of the potential relevance to the S&T developments of national security interest. Even some universities with high level advanced research on cybersecurity can have appallingly bad cybersecurity practices in their overall IT management. Universities in many countries are a unique form of private corporation. In China, they are very much "state-led" organizations but in principle are not government agencies. The norm globally appears to be that cyber security in universities is quite weak, and this appears to be especially the case in China, since it was this sector that was among the hardest hit in that country by the WannaCry ransomware pandemic in May 2017. This attack had global reach through automated propagation. Students and some universities reported that they had been affected by Wannacry. Some teaching systems were also affected. But most importantly, the attack coincided with the time for submission of graduation theses. Since the attack depended on a vulnerability in Microsoft Windows for which a patch had been issued, the impact on Chinese universities showed that in general these institutions have bad security practices (either by not installing updates on the day of issue or by using pirated software).

4.7   Conclusion

China may be among world leaders in the cybersecurity of the country's major banks. However, elsewhere in Chinese corporations and other commercial entities, cybersecurity has been something of an afterthought in the rush for profits or in the face of competing priorities. The situation in China may not be worse than in many countries in some sectors (such as civil aviation and the electric grid), but this is small comfort when the level of cyber crime perpetrated against Chinese corporations seems quite high, alongside unusually high losses of personal data reported in the next chapter. Major reforms in organizational culture, and in work force development, will be essential if China is to lift its very patchy performance in corporate cyber security.

References

Anderson R, Barton C, Böhme R, Clayton R, Van Eeten MJ, Levi M, Moore T, Savage S (2013) Measuring the cost of cybercrime. The economics of information security and privacy. Springer, Berlin, Heidelberg, pp 265–300

Armin J, Thompson B, Ariu D, Giacinto G, Roli F, Kijewski P (2015) August. 2020 cybercrime economic costs: no measure no solution. In: 10th international conference on availability, reliability and security (ares). IEEE, pp 701–710

Austin G (2014) Cyber policy in China. Polity, Cambridge

Austin G, Cappon E, McConnell B, Kostyuk N (2014) A measure of restraint in cyberspace: reducing risk to civilian nuclear assets. EastWest Institute, New York/Brussels/Moscow. https://www.eastwest.ngo/sites/default/files/ideas-files/munich2014.pdf

BoC (2016) Consolidate cybersecurity barriers to enhance financial services. In Chinese. http://pic.bankofchina.com/bocappd/csr/201605/P020160521681204387049.pdf

Brunekrceft G, Luhmann T, Menz T, Muller S, Recknagel P (eds) (2015) Regulatory pathways for smart grid development in China. Springer, Berlin

CAICT (2017) White paper on the cybersecurity industry. In Chinese. Chinese Academy of Information and Communications Technologies. http://www.caict.ac.cn/kxyj/qwfb/bps/201709/P020170919308653198647.pdf

CAS (2014) Possible major S&T breakthroughs in China over the next decade. Bull Chin Acad Sci 28(1):62–105. http://english.cas.cn/bcas/2014_1/201411/P020141121529840357394.pdf

CBRC (2013) China banking regulatory commission annual report 2012. http://www.cbrc.gov.cn/chinese/files/2013/4CF24B3E79704CEA85D330A7CC18CD7D.pdf

CBRC (2015) China banking regulatory commission annual report 2014. Part 1. http://www.cbrc.gov.cn/chinese/files/2015/0F19960DD41D4206A246251A7225773E.pdf

CBRC (2016) China banking regulatory commission annual report 2015. http://www.cbrc.gov.cn/chinese/files/2016/6C1DEC063D6442B289B7C24F662D2E52.pdf

CCTV (2016) State grid exposes data of tens of millions of clients! State power emergency response. In Chinese. South Metropolis Daily. 14 Dec 2016. http://www.xzbu.com/6/view-6614435.htm

Charlton A (2013) Angry Bitcoin users blamed for China Central Bank Cyber Attack. Int Business Times. 19 Dec 2013. http://www.ibtimes.co.uk/angry-bitcoin-users-blamed-china-central-bank-cyber-attack-1429648

Chen L, Liu X, Zhang T, Wang Y (2013) The research on information architecture and security protection of smart grid. Appl Mech Mat 421:541–545. http://citeseerx.ist.psu.edu/viewdoc/download?doi=10.1.1.1008.6294&rep=rep1&type=pdf

China Eastern (2017) 2016 China Eastern Airlines corporate social responsibility report. http://en.ceair.com/upload/2017/5/51132788.pdf

CNN (2015) Cyber Criminals Attack Bank of China for Bitcoin Ransom. 20 May 2015. https://www.ccn.com/cyber-criminals-attack-bank-china-bitcoin-ransom/

CNNIC (2017) Statistical report on internet development in China. Jan 2017. https://cnnic.com.cn/IDR/ReportDownloads/201706/P020170608523740585924.pdf

Gartner (2017a) Gartner survey finds CIOs in China preparing for a digital ecosystem surge. Press Release. 22 Mar 2017. http://www.gartner.com/newsroom/id/3650317

Gartner (2017b) Gartner says worldwide information security spending will grow 7 percent to reach $86.4 billion in 2017. Press Release. 16 Aug 2017. https://www.gartner.com/newsroom/id/3784965

Hathaway M (2103) Cyber readiness index 1.0. Hathaway Global Strategies LLC, Great Falls, VA. http://www.belfercenter.org/sites/default/files/legacy/files/cyber-readiness-index-1point0.pdf ICAO 2017

ICAO (2016) Resolutions adopted by the assembly. 39th session. Montreal. 27 Sept to 6 Oct 2016. Resolution 39–19. Addressing cyber security in civil aviation. Provisional edition. Oct 2016, pp 91–93. https://www.icao.int/Meetings/a39/Documents/Resolutions/a39_res_prov_en.pdf

ICAO (2017) Declaration on cyber security in civil aviation, Dubai, 6 Apr 2017. https://www.icao.int/Meetings/CYBER2017/Documents/Draft%20Dubai%20DECLARATION%20ON%20CYBERSECURITY%20IN%20CIVIL%20AVIATION_10%20March%202017.pdf

Li DC (2015) Online security performances and information security disclosures. J Comput Inf Syst 55(2):20–28

Lin R, Xie Z, Wang X, Wei J (2016) Institutional pressures, legitimation of information security and organizational performance: an empirical study on China's firms. Manage World 2:122–188

Lou X, Fu G, Gong W, Liang Y, Chen Y (2016) Financial institutions: how far are you from the cyber security law? King & Wood Mallesons. http://www.kwm.com/en/knowledge/insights/financial-institutions-how-far-are-you-from-the-cyber-security-law-20161114

McGuire M, Dowling S (2013) Cyber crime: a review of the evidence. Summary of key findings and implications. Home office research report. http://www.justiceacademy.org/iShare/Library-UK/horr75-chap1.pdf

Qihoo 360 (2017a) Analysis report on security vulnerabilities of Chinese websites 2016. In Chinese. 5 Jan 2017. http://zt.360.cn/1101061855.php?dtid=1101062368&did=210133742

Qihoo 360 (2017b) 2016 internet security report. In Chinese. 12 Feb 2017. 209 pp. http://zt.360.cn/1101061855.php?dtid=1101062514&did=490278985

Ran F, Huang H, Ma J, Xu M (2012) Analysis of information encryption on electric communication network. In: Xiao T, Zhang L, Fei M (eds) AsiaSim 2012. Communications in computer and information science, vol 324. Springer, Berlin, Heidelberg, pp 143–150. https://link.springer.com/chapter/10.1007/978-3-642-34390-2_17

Reuters (2014) China pushing banks to drop IBM servers in hacking dispute: report. 27 May 2014. https://www.reuters.com/article/us-ibm-china/china-pushing-banks-to-drop-ibm-servers-in-hacking-dispute-report-idUSKBN0E70S620140527

Reuters (2017) Security experts struggle in search for WannaCry clues—Reuters. 19 May 2017. https://www.finextra.com/newsarticle/30596/security-experts-struggle-in-search-for-wannacry-clues—reuters/transaction

SGCC (2011) SGCC Held integrated information system data disaster recovery drilling. State Grid News. 21 Nov 2011. http://www.sgcc.com.cn/ywlm/mediacenter/corporatenews/12/261468.html

Shen H, Li M, Li Z (2016) An analysis of power grid enterprises' information security system under cloud environment. In: International conference on advanced electronic science and technology (AEST 2016), pp 759–764. www.download.atlantis-press.com/php/download_paper.php?id=25864509

Soo Z (2016) China, Hong Kong firms face highest level of cybersecurity risk, says South China Morning Post. 29 Nov 2016. http://www.scmp.com/tech/article/2050174/china-hong-kong-firms-face-highest-level-cybersecurity-risks-says-pwc

State Grid (2017) Establish a correct cyber-security concept and build a secure line of defence. In Chinese. State Grid Corporation. 2 May 2017. http://www.cec.org.cn/zdlhuiyuandongtai/dianwang/2017-05-02/167735.html

Tang M, Li M, Zhang T (2016) The impacts of organizational culture on information security culture: a case study. Inf Technol Manage 17(2):179–186. http://jtp.cnki.net/bilingual/detail/html/GLSJ201602014

TNO (2012) Cost of cyber crime largely met by business. http://www.tno.nl/content.cfm?context=overtno&content=nieuwsbericht&laag1=37&laag2=69&item_id=2012-04-10%2011:37:10.0&Taal=2

USITO (2017) CAAC Drafted new security measures in line with CSL. http://www.usito.org/news/caac-drafted-new-secuirty-measures-line-csl

Warner R, Sloan RH (2016) Defending our data: the need for information we do not have. 29 July 2016. https://ssrn.com/abstract=2816010

Wolf J, Lehr W (2017) Degrees of ignorance about the costs of data breaches: what policymakers can and can't do about the lack of good empirical data. 31 Mar 2017. https://ssrn.com/abstract=2943867

Xinhua (2016) China's civil aviation and cyber security companies join forces to tackle the challenges of cyberspace security. In Chinese. 17 Aug 2017. http://news.xinhuanet.com/air/2016-08/17/c_129236824.htm

Xinhua (2017) Strengthening the construction of the 'Blue Team'. Improving the information security level of the power grid. In Chinese. 25 Aug 2017. http://www.cq.xinhuanet.com/2017-08/25/c_1121539075.htm

Young J (2016) China's Central Bank hires Blockchain experts to launch Cryptocurrency. Coin Telegraph, 16 Nov 2016. https://cointelegraph.com/news/chinas-central-bank-hires-blockchain-experts-to-launch-cryptocurrency

Zeng Z, Yang K, Zhang Y and Zhou P (2013) Increasing employees' awareness and enhancing motivation in e-government security behavior management. In: Fourth international conference on digital manufacturing and automation (ICDMA), pp 684–687

Zheng S (2017) Ditch your boarding pass. South China Morning Post. 30 June 2017. http://www.scmp.com/news/china/article/2100646/china-southern-airlines-countrys-first-carrier-use-facial-recognition

Ziser KK (2017) China telecom global launches SD-WAN service with versa networks. Light Reading. 15 Sept 2017. http://www.lightreading.com/carrier-sdn/sd-wan/china-telecom-global-launches-sd-wan-service-with-versa-networks/d/d-id/736388?_mc=RSS_LR_EDT

Chapter 5
Cyber Insecurity of Chinese Citizens

Abstract The chapter restates the well known fact that China's government demands a near absolute right to monitor online activity of its citizens. It is introduced by results of a 2017 report on citizen security in cyberspace by major cities in China. It is then organized around four themes: defenses against cyber crime; the cybersecurity confidence of the citizens; protection of political privacy, including in connection with advanced artificial intelligence; and other personal privacy, especially in the light of the emerging social credit system.

5.1 i-Dictatorship

Chinese leaders see the citizenry either as facilitators of state power, as a troublesome or unruly mass that cannot be trusted with electoral democracy, or in some cases as the enemy. The government does allow "civil society" and community organizing in non-sensitive areas of activity, but almost all of it is closely supervised or monitored. As is widely known, Chinese citizens are the targets of mass surveillance in cyberspace from their own government on a scale unequalled elsewhere in the world. One corollary of this is that its citizens have very poor cyber defenses against the state. At the same time, they must grapple with the same sort of cyber crime and online threats, including breaches of personal data, that citizens of other countries face. A large slice of global criminal activity is generated in China, and therefore Chinese citizens are also victims of their more criminally-minded compatriots. Given that the country as a whole has poorly developed cybersecurity awareness, a large skills deficit in the field, and only modest penetration by the most expert non-Chinese cybersecurity firms, we might reasonably expect the actualities of insecurity in cyberspace for Chinese citizens to be quite immense.

The Chinese government corrupts the security of its citizens in cyberspace the moment they go online. It was supported in this effort at the start by a number of foreign corporations, especially Cisco (Barmé and Ye 1997). This remains the case today, though the band of foreign corporations has grown and been joined by equally willing foreign governments, such as Singapore and Russia, and willing

© The Author(s) 2018
G. Austin, *Cybersecurity in China*, SpringerBriefs in Cybersecurity,
https://doi.org/10.1007/978-3-319-68436-9_5

researchers from foreign universities, and even several foreign religious organizations who promote content filtering and monitoring of personal use of content deemed unsatisfactory.

Since 1997, the Chinese government has wanted its citizens to believe that the state is omnipotent in cyberspace and this intent has been sustained to date, taking on new forms: what we can call *i*-dictatorship (Austin 2014: 62–74). One of the newer forms is the doctrine of internet sovereignty, the main purpose of which is to intimidate the country's citizens to believe that government monitoring and response capacities are near omnipotent. Western commentators who hold up Chinese claims of internet sovereignty as desirable or possible are actually unwitting (or intending) accomplices of the Chinese propaganda machine. The preference of the leaders is that "ordinary people should be allowed full play in safeguarding the country's cybersecurity" in cyberspace (Cao 2017a, b), as coopted supervisors acting on behalf of the CCP.

5.2 National Variations

Readers who might question the severity of these political arguments may like to reflect on an interesting result from a May 2017 report jointly undertaken by four laboratories[1] in China (Ceprei Laboratory 2017). This study analysed certain measures of the cybersecurity situation in 36 cities across three user groups: government, corporations, and individuals. Each of the three user groups in each of the 36 cities was accorded an index score, and a total index score was also estimated for each city. The city of Lhasa, which was ranked 33rd of 36 for government security and 30th for corporate security, was ranked 4th in the country for individual cyber security. In the category of individual cyber security, Shanghai ranked 33rd and Beijing came in at 35th. Several of the characteristics of individual security that were assessed in this report addressed fraud cases or attempts, and for this reason the index may reflect more the wealth of the city in which the individuals live rather than any deficiencies in the security profile of the user. However, Lhasa scored very well on PC Trojan kill, mobile Trojan kill, genuine software use, defences against phishing, and on lower levels of vulnerabilities detected. It seems that while the government agencies and corporate entities in Lhasa have very low cybersecurity, the user population of individual citizens has very high security awareness levels. While not conclusive proof, this would seem to reflect the very high degree of suspicion by Lhasa's citizens about their vulnerability to cyber threats from the government. This assessment is borne out in part by the higher ranking for Urumqi for individual security (17th) versus corporate security (36th) and government security (25th). Urumqi is the site of high internal security scrutiny against the Uighur population.

---

[1]National Engineering Laboratory of Big Data Collaborative Security Technology, National Engineering Laboratory of Big Data Application Technology to Improve Governance, China Ceprei Laboratories, Key Laboratory of Big Data Strategy.

5.3   Defences Against Cyber Crime

All unauthorized access to a citizen's computer systems and data is a potential threat to their security, and the majority of such actions are criminal acts in China, whether under specific cyber-related legislation or general provisions of the criminal law (Li 2015: 186–90). Thus, even before looking at the more newsworthy aspects of cyber crime, such as fraud, theft or blackmail, there needs to be a full accounting of the scale of criminal assault on the systems and data of individual citizens in China. In addition, because of the criminal character assigned by the Chinese authorities to a wide range of political and personal content that would not be regarded as criminal in liberal democracies, the cyber crime picture in the country is unique.

The picture painted for 2016 by Qihoo 360 (2017a: i–iv), based on its own monitoring or reporting to its help centre, makes bleak reading:

- 190 million malware items found on PCs (down 47% on 2015)
- 14 million malware attacks in Android platforms, 74% of which accessed and used victims' funds
- 70,000 mobile phone extortion attempts
- 197 million phishing attacks intercepted
- average life cycle of fraud calls with similar operating method was about 57.6 days, with a continuous active period of about 7.6 days
- 17.35 billion spam messages, of which 4.2% were "illegal", 2.8% were scams
- security vulnerabilities in 99.99% of Android phones, largely because of failure to update on the part of manufacturers and/or users
- 20,000 users nationwide reported monetary losses of 195 million RMB, accounting for 9471 yuan per person, and 56% of the victims proactively transferred money to the fraudsters
- people born after 1990s were the largest group of victims, but older people fell for more costly scams
- illegal or unlawful use (co-option) of smart cameras increased.

CNNIC reported in 2017 that for the previous year that some 71% of netizens were subject to security incidents and offered the following breakdown of types of cyber-security incidents encountered by internet users (CNNIC 2017: 16)

- Fraud: 39%
- Malware: 36%
- Stolen accounts or codes: 34%
- Leaked personal information: 33%
- Others: 30%.

There is little reliable evidence to enable a confident assessment of citizen defences to cyber crime. The Ceprei Laboratories (2017) analysis referred to above shows (as might be expected) wide national variation in the cybersecurity preparedness of individuals between 36 cities across eight categories of observed incident or behavior.

That the weakest performing (the bottom six cities) are major population, business and technology centres (Tianjin, Qingdao, Shanghai, Shenzhen, Beijing, and Chongqing) may be interpreted as a likely sign that cybersecurity at the individual level across the country is very weak indeed, not least because the analysis did not include another 60 cities in China with population over one million or any rural areas (another 500 million people at least).

The following sub-sections provide some glimpses into several different types of cyber crime in China: ransomware, fraud and data leaks. The last one mentioned —data leaks—is not always a criminal act, but as reported below, is the stepping stone to much criminal activity in cyber space.

5.3.1 Ransomware

Qihoo 360 (2017a: 2–3) has identified the rise in malware in PCs as particularly problematic, as did the China CERT (CNCERT/CC 2017: 20). The Qihoo report talked of a doubling of use of viruses and, citing IBM Security, an estimated explosion of this technique globally by a factor of ten in 2017. It said that 113 new ransomware attacks had been detected, involving 167,000 samples, and affecting "at least 5 million" users across the country. The Qihoo report identified three main vectors of attack: Trojans, email (phishing), and server intrusion. There were two large scale scams: one in April–May linked to pornographic websites abroad co-opted by hackers and one in September–October caused by a large financial services platform in China penetrated by hackers. In spite of a sharp decline between 2010 and 2014 of trojan attacks on websites in China, after 2015 this tactic resurfaced (Qihoo 360 2017a: 3), though outside of those two peaks just mentioned, that data shows only low numbers. The main causes for the increase were the leak of Hacking Team data, increase in vulnerabilities in Adobe Flash (with longer repair times), and a return to favour among criminal gangs of this technique. In 2016, the main Trojan attack vectors were vulnerabilities in Internet Explorer (88.5% of attacks) and Adobe Flash (11.5%). Individual users accounted for 81.1% of targets and they are more vulnerable because of lack of security knowledge.

According to the survey data, men were the most vulnerable to the ransomware, accounting for 93.2% of the total (Qihoo 360 2017b: 7). (This statistic varies slightly in other versions of the report apart from the one referenced.) The survey also revealed that the main reason why male victims were infected with blackmailers was by browsing unfamiliar web pages. Among the victims who sought help from 360 Internet Security Centre, only 37.6% were individual users (as opposed to corporations), in spite of accounting for 81.1% of monitored attempts. Some 42.6% of the victims did not know how they were infected, while 21.8% of the victims were infected while browsing unfamiliar web pages; 11.9% infected when downloading software; 8.9% of the victims were active clicking on the attachment of the email to catch the virus; 5% of the victims were infected through a USB flash drive; 3.5% of the victims are on port 3389 (remote control port with the Windows system) (Qihoo 360 2017b: 8).

The sample survey showed that among the victims who asked for help from 360, only 11.9% of victims had paid a ransom to recover their files (Qihoo 360 2017b: 10). Among those, 58.4% paid using the Taobao e-commerce platform (with over 90% recovering their encrypted files); and 33.3% of the victims were asked to exchange bitcoin (with only 50% of those recovering their files.

In its 2016 report, Antiy (2017: 19) referred back to its assessment one year earlier that ransomware would become the most direct threat globally to individual users and enterprise customers and concluded that this assessment had been proven through the course of the subsequent year. Moreover, Antiy said, the criminals had become more adept at targeting higher value files to encrypt and use for ransom.

5.3.2 Fraud

In 2016, the government's incident reporting platform recorded 20,623 cases of online fraud from users in the country, involving more than 195 million RMB, with an average loss of just under 10,000 RMB per incident (Qihoo 360 2017b: 59–64). Compared with 2015, the number of reported online fraud cases declined by 17.1%, but the per incident average loss almost doubled (+85.5%). PC users accounted for around two thirds of the reported cases in 2016, with marginally higher losses per incident, while mobile users accounted for under a third of cases and reported smaller average losses (under 7000 RMB) per incident.

The fraud cases were heavily concentrated in financial management, such as purchase of online financial products (37% by value), gambling (15%), part-time fraud (11%) and identity theft (7%). There were several attack routes: voluntary transfer (56%), fake phishing sites (35%); unauthorized access based on voluntary handover of passwords or other privacy information on phishing websites (5%); and Trojan software (2%). Guangdong was the hotspot among 31 provinces and municipalities, accounting for just over 10% of the cases. Men were more susceptible (74% of cases) but there was no significant difference in average losses between men and women.

People born after 1990 accounted for 42% of the victims. The data shows that online fraud is increasingly effective among younger people, especially teenagers, but it was older people on average who lost more money, with the amount of average loss increasing with age cohort. In August 2017, Chinese authorities convicted seven criminals involved in defrauding a young student, Xu Yuyu, in 2016 of the money raised by her family to pay her tuition fees. The case was notorious because Xu died of a heart attack in the presence of her father after reporting the case to police.

In 2016, in their malware detection, disposal and destruction campaign, CNCERT focused on message communication, and Android malwares like Album (which fool users into download of export functions) and on porn-embedded software equipped with the property of malicious charging and malicious communication (CNCERT/CC 2017: 5). A total of 47,316 such malicious programs were found throughout the year. More than 1.01 million users were infected. The number of domain names used to spread malicious programs was 6045. There were 7645 malicious email accounts used to receive users' messages and address books. There were 6616 malicious phone numbers used to receive users' messages. There were 2.22 million emails used to leak users' message and address books.

5.3.3 Data Leaks

In 2017, CNCERT/CC reported that "the threat of data leakage has intensified" in the previous year (19). Both domestic and foreign website data and personal information leakage events were frequent. The CERT also reported (24) that "A large number of actual cases and studies have shown that personal information disclosure has become an online fraud booster. More than half of fraud cases were based initially on data leakage (ticket signing, shopping refund, pretending to be a public prosecutor, pretending to be acquaintances). According to the Internet Society of China, about 51% of Internet users leaked their personal information in the process of online shopping, and 84% encountered harassment or loss of money" (cited by Qihoo 2017b: 174). The scale of economic loss due to personal information leakage was estimated by ISOC to have reached 91.5 billion yuan (China Daily 2016). In December 2016, the Guangzhou-based *Southern Metropolis Daily* reported that it only cost 700 yuan to buy 11 items of personal privacy information for particular individuals, including personal flight record, deposit, check-in record, mobile phone real time positioning information, and mobile phone call records (Rao and Li 2016). According to Antiy (2017), a large number of personal information leaks in 2016 were caused mainly by two reasons: vulnerabilities in websites that are hacked and criminal sale by insiders of the organizations concerned. The CNCERT Annual Report for 2016 provides some indication fo the scale of data leaks for that year, in the indicative list in Table 5.1.

According to CNCERT/CC (2017: 19), the Ministry of Public Security prosecuted more than 1800 cases of personal information theft, in which they found more than 30 billion separate records of personal information. According to Qihoo's data, criminals stealing personal information from mobile phones were most interested in text content (67%), followed by mobile banking information (35%), supporting theft of mobile phones (10%), and theft of mobile phone records (4%) (Qihoo 2017b: i).

5.4 Cybersecurity Confidence: High and Low Awareness

One measure of individual cyber security in China is the confidence that users have in the security of the internet in terms of its protection from crime and from breaches of privacy. According to CNNIC (2017: 105–106), the annual survey of users in 2016 found that 38.8% considered China's network environment safe or very safe, and 20.3% regarded it as unsafe or very unsafe. The report also noted that 70.5% of those surveyed had encountered cyber security incidents. This data seems to suggest

Table 5.1 List of notable cases of losses of personal data in 2016

Institution	Estimated loss (no. of records)
Population and Family Planning Commission in one province	70 million
Financial system	36 million
Health system in one province	18 million
Logistics system	16 million
Hotel	9 million
Shopping website	8 million
Airline company	6 million
Insurance system	5.5 million
Communication system	3 million
A communication company system	3 million
Social Security Bureau in one city	2.7 million
Bank	2.57 million
Social Security system in one province	2.13 million
Postal communications company website	1.8 million
Government affairs office in one province	1.71 million
An engineering quality supervision bureau in a department	1.3 million
Bank website	1 million
Online office of Social Security in one city: 1,000,000+	1 million
Tourism website	1 million
Mobile application of Department of Energy	830,000
Communication company website	810,000
Service website	570,000
A university system in a city	500,000
Children's vaccination registration information in one city	200,000
Total	193 million

slightly more confidence than among users in the United States, though the environment and the survey questions are quite different. For example, according to Pew (Olmstead and Smith Olmstead and Smith 2017) 64% of Americans surveyed "have personally experienced a major data breach", "relatively large shares of the public lack trust in key institutions—especially the federal government and social media sites—to protect their personal information", and 35% "have received notices that some type of sensitive information (like an account number) had been compromised".

China has invested quite heavily in cybersecurity awareness campaigns but as suggested earlier, the government objectives have several themes: take personal responsibility for knowing and acting; trust the government; and report your work colleagues or others for breaches of their social contract on content security.

The Xi announcement in February 2014 on cyber power and the importance of national cyber security appears to have been an additional catalyst. The country launched the first national cybersecurity awareness week in November 2014 and the second was held in June 2015 to improve education about teenagers' safety (Hui and Tan 2016: 13). The second iteration had separate theme days, including Finance, Telecoms, Government Affairs, Science and Technology, Rule of Law, and Teenagers. The fourth iteration was held in September 2017. The CAC, the Ministry of Education, the Communist Youth League, and the Internet Society have been especially active participants in these campaigns directed at young people. The Internet Development Foundation, launched in August 2015 and set up for similar purposes, was the first public fundraising foundation in China's Internet sector (Hui and Tan 2016: 13). Its aims include educating more Internet talent, promoting China's international interests in cyberspace, and contributing to a "safe, peaceful, open and cooperative cyberspace". Various skill contests are now held across China (in schools, workplaces, and in civil society) and at an escalating pace, with more than 100 in 2015 alone (Hui and Tan 2016: 13). Chines individuals and teams regularly participate in international competitions.

For a number of years, China has taken a leadership role in advocacy for online protection of children, though this has a political dimension of building online social-ist culture as well as the protection from classic threats existing around the world. Moreover, there are few reliable and consistent statistics in China on the number of cyber-enabled crimes against children, much less on the number of investigations and convictions.

It is China's practice to occasionally announce the busting of gangs or the con-viction of perpetrators of cyber crime but little policy analysis in the public domain that might allow a judgment about how effective awareness and prevention programs are or can be. Criminology in general has had a slow rebirth beginning in earnest in 1992, and by 2004 two scholars assessed that "the application of modern methods of social science research as applied to a whole range of criminal justice issues is in its infancy" (Broadhurst and Liu 2004: 4). Problems identified then, such as official rectitude toward release of data, especially where crime reveals failures of gover-nance, or where there are political sensitivities, are still visible today. A 2017 study assessed that "criminological research in China is progressing and becoming more sophisticated" but that "crime and justice research in China remains underdevel-oped by Western standards" (Liu and Li 2017: 449). As another author notes, crime statistics can often be manipulated for political purposes and "should be more fully understood as part of a legitimization apparatus in China" (Xu 2017: 155). There has been a reorientation toward empirical studies to bring Chinese criminology more into line with international social science practice (Hebenton and Jou 2018: 386). These authors report that there are three main focal points of such research: the Crime Prevention Institute in the Ministry of Justice, the MPS Public Security Institute, and university-based research.

Criminological research in China on cyber crime may be something of an exception, with data from CNKI showing clear spikes in academic publications in 2004 and 2010. These have been mainly journal articles, and remain small in number relative to the size of the country's research community and policing budgets and taking into account the diversity of possible subjects under the rubric of cyber crime.

There have been only a handful of doctoral dissertations on social science aspects of cyber crime in the past twenty years.

Academic research on security awareness in China and how it impacts behavior, crime outcomes, and consumer attitudes is in its infancy even as threats and platforms proliferate. The general situation appears to be rather well captured in one study of internet banking (Zhu 2015), which found that the general level of consumer awareness of security issues and responses in that sector remained low. It must be remarked however that assessment of such issues is not a simple measurement problem but a highly complex one that demands polycontextual approaches, that include factors such as security perceptions, coping strategies, cohort age, gender, and professed approach to citizenship and governance (or espoused culture) (Chen and Zahedi 2016). It is also possible to take many different analytical perspectives on the security practices of even a single cohort (e.g. Beijing adult males) in a single security environment (e.g. e-auctions). The construction of any sort of judgement about national level preparedness of individual citizens for the diversity of cyber security challenges is fraught with challenges, and the best analytic work appears to be that which is limited to specific cohorts, sectors and threat types.

5.5 Political Privacy and Advanced AI

In China, a citizen's political views are not regarded as a private affair. In the practice of the CCP's supervision of political life, including in cyber space, there is no presumption of freedom of conscience or freedom of belief. Since the earliest days of the access by ordinary citizens to the Internet in China in the mid-1990s, there has been a government priority to monitor and control online expressions of political opinion. This is reflected in a vast number of Chinese official documents and has been described in numerous sources, summarized in my own work under the term of *i*-dictatorship (mentioned above). The intensity of government interest in particular dissenting or unacceptable opinions, or in particularly individual people, can change over time.

The enduring situation was captured very well by a Chinese official interviewed by *Wired* magazine as early as 1997, at a time when 86% of the country's citizens had never touched a computer. The official said: "People are used to being wary, and the general sense that you are under surveillance acts as a disincentive. The key to controlling the Net in China is in managing people, and this is a process that begins the moment you purchase a modem" (Barmé and Ye 1997). The official described the use of key words as a filter to block unacceptable material, but also noted that many materials were blocked simply because of where they came from.

Since this phenomenon has been widely studied, as summarized in Austin (2014: 62–74) this section seeks merely to provide some additional reference points for the period since the Xi cyber power announcement in 2014. Various official documents on new approaches have been published since then, and mentioned above, such as the Cyber Security Law and the National Cyber Security Strategy. Chapter Two also mentioned a renewed emphasis on university education for content control.

One of the most important innovations since 2014 has been the establishment by the MPS of over 1000 cyber security police units, with variegated capabilities according to their mission (Wang 2017). For example, Level One units have been stationed inside thirteen major website companies, including Baidu, TenCent and Sina. The report on this development also observed that since the units had been set up beginning in August 2015, the MPS had "shut down 250,000 illegal online columns, groups, and shops, and have closed 470,000 unlawful accounts". It also commented that these cyber police units had been able to greatly expand the supply of "criminal" intelligence. (The term criminal here includes reference to political offences.) The situation prior to the establishment of these units was that internet companies were required, under a policy of "social control", to conduct the monitoring themselves. The leadership apparently judged that their efforts were proving inadequate.

Another notable turn was a crack-down in 2017 on VPN use, making it illegal for anyone in China to use this software without registering it with the government. The VPN allows users to access material without exposing it to the internet monitors. Hong Kong's South China Morning Post reported on the possible impacts on China's research community with a sub-headline: "Beijing risks a brain drain and undermining international collaborations by cutting off academics reliant on virtual private networks" (Zheng 2017).

The 2017 crack-down on citizen access to the Internet also saw an intensification of Chinese efforts, that began at least as early as 2002, to ensure real name registration of anyone accessing a user account through an ISP. In August 2017, the government announced new regulations to be enforced beginning 1 October, forbidding anonymous access to the internet and making ISPs responsible for verifying genuine identities of all account holders (Xinhuanet 2017).

The period after February 2014 also saw a massive increase in interest by the MPS in new applications of artificial intelligence, including facial recognition of people on public transport and in crowded places, as well as near-instantaneous recognition and identification of the license plates of cars (Li 2017). In July 2017, the government announced a radical plan to lead the world in AI research by 2030 (Lee 2017). The plan would see a 100-fold increase by China in the field compared with its 2012–2016 spend. Several months later, the Chair of the CPLC, Meng Jianzhu, was quoted as saying: "Artificial intelligence can complete tasks with a precision and speed unmatchable by humans, and will drastically improve the predictability, accuracy and efficiency of social management" (Gan 2017). The latter report cited a Germany-based Chinese scholar, Zhou Zunyou, saying that "China has no specific data protection law. The government can use personal data in any way they like, which could pose a huge threat to its citizens' privacy."

5.6 Other Forms of Cyber Privacy: Social Credit

It may be tempting to downplay the above analysis about lack of political privacy in China by reference to the very genuine efforts in the country, including from the CCP, to promote a high level standard for privacy in general and for protection of citizens' privacy online (see Austin 2014: 80–83). There are three main reasons to avoid such a temptation. The first is that the scale of data breaches in China is very high. Second, the standards of government and corporate cyber security are, on average, low. Both of these issues have been flagged. The third reason is far more serious: the leaders have announced the biggest anti-privacy measure ever seen in history, an automated 'social credit system' to be mandatory by 2020.

The Social Credit System was flagged as a priority in the 18th Party Congress in November 2012 that saw Xi Jinping take over as General Secretary. It was further developed in the third Central Committee plenum one year later (as part of the 60-point reform plan to be the foundation of Xi's stewardship of the country) in the following terms: "establish and complete a social credit system, commend sincerity and punish insincerity". The full import of the plan, and its deeply intrusive character to be based on cyber technologies, was only fully revealed in the guidance issued in 2014, where the central plan of the policy was revealed as "Sincerity in government affairs" (State Council 2014).

The system is intended to promote and document the social credit of:

- every Chinese citizen (in their roles as private citizen, employee, or social actor) as a vehicle for decision-making about that person's access to employment, social services, health services, education, public transport
- every Chinese governmental official in all fields, but especially in the criminal justice system, as a vehicle for decision-making about their work performance and employment future
- corporate actors as a vehicle for decision-making about their access to bank loans, conformity with their tax obligations, and even the right to operate.

The system is to be built on the assignment of scores for different actions and omissions, to be moderated by algorithms and facilitated by a country-wide network of local information systems. A whole new profession of social credit evaluators will be constructed. They will need IT support in the form of:

- sectoral credit information systems and databases that are integrated and secure
- local information systems to accelerate the integration of government affairs information, including through information sharing platforms to allow access for enterprises, individuals and social credit investigation bodies
- credit investigation systems to "collect, arrange, store and process credit information of enterprise and undertaking work units and other social organizations, as well as individuals"
- uniform credit investigation platforms in the financial sector
- credit information security management systems.

The outline makes a plea for protecting the rights and privacy of the subjects of the social credit system, but this is clearly and aspirational statement and one that few Chinese citizens would take seriously. The combination of advanced artificial intelligence with the social credit system is definitely power-enhancing, potentially at least. As Hoffman (2017) observes, it is a method for "automating the 'carrot' and 'stick' processes that ensure the Party-state's power".

That said, we cannot dismiss the Chinese government's interest in a social credit system in the broad as exclusively a sign of an authoritarian state, as Backer (2017) has argued in his excellent overview of social credit as a global phenomenon. But we have to examine its planning and implementation in detail to assess its impact. Backer warns that any social credit system will meet resistance (8). One Chinese scholar argues for consideration of the possibility that the use of massive digital data "might backfire against the authoritarian regime" as it may change power balances if used by competing political factions against each other (Zheng 2016: 1444). One joint study of government censorship policy of mass media raises a consideration that is even more apt for the prospective social credit system: "the increasing complexity of the CCP's information-control apparatus may be unsustainable" (Clark and Zhang 2017: 786).

5.7 Conclusion

Analysis of the cybersecurity of Chinese citizens, now and in prospect, illuminates better than most lines of inquiry the true character of the cybersecurity ecosystem in the country as a whole. It is an ecosystem intended to serve the interests of an authoritarian dictatorship and not the interests of individual citizens. This will be increasingly evident as the Chinese state partners with foreign corporations in advanced artificial intelligence applications in support of both its monitoring/censorship polices and development of its social credit system, relying on accumulation of personal data of the citizenry. Successes in battling cyber crime in China probably arise more from the authoritarian instincts of the state than from any strong commitment to protection of the citizenry. While the government is committed at a rhetorical level to eliminating all forms of crime to the extent possible, it has not invested significantly in the criminological research on cyberspace affairs that might begin to inform better and more effective policy in this area. In fact, our knowledge of the cybersecurity of Chinese citizens is based on an understanding of the broad contours of the ecosystem, occasional surveys and a handful of case studies, and not on comprehensive detailed research by Chinese and/or foreign scholars.

References

Antiy (2017) Review and prospects for cybersecurity threats in 2016. In Chinese. http://www.antiy. cn/report/2016_Antiy_Annual_Security_Report.html#2

Austin G (2014) Cyber policy in China. Polity, Cambridge

Backer LC (2017) Measurement, assessment and reward: the challenges of building institutionalized social credit and rating systems in China and in the West. In: Proceedings of the Chinese social credit system, Shanghai Jiaotong University, 23 Sept 2017. https://ssrn.com/abstract=3040624

Barmé G, Ye S (1997) The great firewall of China. Wired. 1 June 1997. https://www.wired.com/1997/06/china-3/

Broadhurst R, Liu J (2004) Introduction: crime, law and criminology in China. In: Aust NZ J Criminol 37(1_suppl):1–12

Cao S (2017a) China to mobilize net users for cyber security. Global Times. 13 Sept 2017. http://www.globaltimes.cn/content/1066130.shtml

Cao S (2017b) Telecom fraudsters plead guilty in student death case. China Daily. 28 June 2017. http://www.chinadaily.com.cn/china/2017-06/28/content_29911556.htm

Ceprei Laboratory (2017) Big data city network security index report. 360 cyber security center. http://zt.360.cn/dl.php?filename=%E5%A4%A7%E6%95%B0%E6%8D%AE%E5%9F%8E% E5%B8%82%E7%BD%91%E7%BB%9C%E5%AE%89%E5%85%A8%E6%8C%87%E6% 95%B0%E6%8A%A5%E5%91%8A.pdf

Chen Y, Zahedi FM (2016) Individuals' internet security perceptions and behaviors: polycontextual contrasts between the United States and China. MIS Q 40(1):205–225

China Daily (2016) Online infringements cost $13.8b a year. 24 June 2016. http://www.chinadaily. com.cn/business/2016-06/24/content_25841504.htm

Clark C, Zhang L (2017) Grass-mud horse: Luhmannian systems theory and internet censorship in China. Kybernetes 36(5):768–801

CNNIC (2017) Statistical report on internet development in China 2017. https://cnnic.com.cn/IDR/ ReportDownloads/201706/P020170608523740585924.pdf

Gan N (2017) China's security chief calls for greater use of AI to predict terrorism, social unrest. South China Morning Post. 21 Sept 2017. http://www.scmp.com/news/china/policies-politics/ article/2112203/china-security-chief-calls-greater-use-ai-predict

Hebenton B, Jou S (2018) Criminology in China. In: The handbook of the history and philosophy of criminology. (ed) Triplett RA. Wiley, 377–391

Hoffman S (2017) Managing the state: social credit, surveillance and the CCP's plan for China. Jamestown Foundation. 17 Dec 2017. https://jamestown.org/program/managing-the-state-social-credit-surveillance-and-the-ccps-plan-for-china/

Hui Z, Tan Q (2016) Cyberspace security in the era of data economy: global and Chinese contexts. In: Hui Z, Tan Q (eds) Annual report on development of cyberspace security in China. Social Sciences Academy Press. Blue Book, Beijing, 1–16

Lee A (2017) World dominance in three steps: China sets out road map to lead in artificial intelligence by 2030. South China Morning Post. 21 July 2017. http://www.scmp.com/tech/enterprises/article/ 2103568/world-dominance-three-steps-china-sets-out-road-map-lead-artificial

Li T (2017) Chinese facial recognition start-up eyes global opportunities beyond public security. South China Morning Post. 22 Nov 2017. http://www.scmp.com/tech/start-ups/article/2121100/ chinese-facial-recognition-start-eyes-global-opportunities-beyond

Liu L, Li JCM (2017) Crime and justice research in China remains underdeveloped by Western standards. J Res Crime Delinquency 54(4):447–453

Olmstead K, Smith A (2017) Americans and cybersecurity. http://www.pewinternet.org/2017/01/ 26/americans-and-cybersecurity/

Qihoo 360 (2017a) Research on Internet fraud trends for 2016. In Chinese. 13 Jan 2017. http://zt. 360.cn/1101061855.php?dtid=1101062366&did=210142130

Qihoo 360 (2017b) 2016 China Internet security report. In Chinese. 12 Feb 2017. http://zt.360.cn/ 1101061855.php?dtid=1101062514&did=490278985

Rao L, Li L (2016) Terror! Nandu reporter 700 yuan to buy the whereabouts of colleagues and other 11 records. Southern Metropolis Daily. In Chinese. 12 Dec 2016. http://epaper.oeeee.com/epaper/A/html/2016-12/12/content_103959.htm?from=timeline&isappinstalled=0&winzoom=1

State Council (2014) State council notice concerning issuance of the planning outline for the construction of a social credit system (2014–2020). GF No. (2014)21. For english version see Planning outline for the construction of a social credit system (2014–2020). China Copyright and Media.https://chinacopyrightandmedia.wordpress.com/2014/06/14/planning-outline-for-the-construction-of-a-social-credit-system-2014-2020/

Wang H (2017) 1116 cyber-security police units have been set up nationwide. Guangming Online, Beijing. In Chinese. http://epaper.gmw.cn/gmrb/html/2017-02/14/nw.D110000gmrb_20170214_2-04.htm

Xinhuanet (2017) China to punish illegal publicity on internet forums. 25 Aug 2017. http://news.xinhuanet.com/english/2017-08/25/c_136554917.htm

Xu J (2017) Legitimization Imperative: the production of crime statistics in Guangzhou, China. Br J Criminol 58(1):155–176

Zheng J (2016) China's date with big data: will it strengthen or threaten authoritarian rule? Int Aff 92(6):1443–1462. https://doi.org/10.1111/1468-2346.12750

Zheng S (2017) VPN crackdown an 'unthinkable' trial by firewall for China's research world. South China Morning Post. 23 July 2017. http://www.scmp.com/news/china/policies-politics/article/2103793/vpn-crackdown-unthinkable-trial-firewall-chinas

Zhu R (2015) An initial study of customer Internet banking security awareness and behavior in China. In: Pacific Asia conference on information systems. PACIS 2015 proceedings, vol 87. http://aisel.aisnet.org/pacis2015/87

Chapter 6
Governmental Cybersecurity

Abstract This chapter looks at how well prepared the Chinese government is to defend its various cyber assets. These range across internal security, leadership security, national defence, and protection of critical information infrastructure. There are competing tensions between the different strands of policy. Moreover, everything that seems largely domestic has an inescapable international dimension, and vice versa.

Chinese leaders face enormous constraints in pursuing their preferred policies in cyberspace. The "Great Firewall of China" is often seen by outsiders as a powerful weapon in the government's arsenal. There is some cause. The march of technology, especially artificial intelligence, is strengthening the coercive potential of the Chinese government in cyberspace, making the Great Firewall more effective in some technical senses. Yet, its strategic utility may be as flawed as that of the original Great Wall. At the same time, other countries and their corporations are not standing still in their efforts to undermine the Great Firewall. Some foreign private sector companies at the cutting edge of technology and with no business stake in China work very hard to develop and sell tools and knowledge that undermine Chinese national security in cyberspace in the broad and, in particular, to weaken the technological foundations of the Great Firewall. This chapter provides several windows on the state of governmental cyber-security in China: a statistical glance, the cybersecurity defence capability of its public security intelligence (PSI) system, protection of critical information infrastructure (CII), and the impact of "internet terror" on Chinese leadership views of cybersecurity.

6.1 A Statistical Glance

As noted above, in-depth study by Chinese and foreign specialists of the state of cybersecurity in China as a public domain activity is a relatively new phenomenon. This is especially the case for cyber practices in government agencies, which provide a more politically sensitive profile than either the cyber defences of corporations and

© The Author(s) 2018
G. Austin, *Cybersecurity in China*, SpringerBriefs in Cybersecurity,
https://doi.org/10.1007/978-3-319-68436-9_6

citizens reviewed in previous chapters. As a baseline, we can take a 2011 assessment by a Chinese specialist that government public-facing networks are "so vulnerable that hackers or computer experts can easily intrude into systems to get the server's user account information and password files, and modify and delete some important data files at discretion" (Luo 2011: 394).

The 2017 Ceprei Labs report provides one of the first reliable sets of indicators for government cybersecurity in 36 cities in China. The results show surprisingly low scores in politically sensitive locations, such as Lhasa (Tibet) and Urumqi (Xinjiang), and wide variations among more developed cities on certain sub-indexes. Table 6.1 shows a sampling of some of the data for ten cities. Note that the higher the score on the sub-index or index, the more secure the assessed target.

6.2   Cyber Protection Ecosystem in China's Security Apparatus

The crowning glory of Chinese cyberspace security is its internal surveillance and intelligence collection system. The Chinese terms for this activity is "public security intelligence" (PSI). China may be the most capable country in the entire world in this regard, and as mentioned in Chap. 1, the country is only a few decades away from having a decidedly Orwellian capability in this field. This section of the chapter uses an understanding of China's PSI capability as a proxy for its ability to defend its governmental cyber systems. The proposition is that the weaker its PSI is in cyber space, the weaker its defences in cyber space are in general.

We know that China deploys large numbers of people in a wide variety of roles to patrol cyberspace and collect intelligence through it. We know that China uses cutting edge technologies for this, including where necessary foreign technologies. We also know that China's technologically-driven, automated censorship and surveillance can be highly effective. The former blocks key words or websites, while the latter tracks key words and websites for alert to items of intelligence interest. The MPS maintains a robust and expanding R&D capability of its own, mostly through its Third Research Institute and its Intelligence Science and Technology Institute (*qíngbào kējì suǒ*). As early as 1994, three of the four MPS research institutes had a heavy focus on information security (Wong 1994: 5–13).

We have less understanding from public domain sources of the quality of cyber defenses of these internal agencies. Can they protect their own cyber secrets? It is far from clear in the public record just where and in what manner electronic data is collected and stored by these agencies, but there is considerable evidence that the security ecosystem for this information is in a rather poorly developed state. If these agencies can't deliver high standards of cybersecurity for themselves, then it is questionable just how effective they can be in supporting information security of other government agencies across the breadth of the country.

Table 6.1 Data on government agencies cyber security in selected cities (Ceprei Labs 2017)

Rank	City	Vulnerabilities detected	Reported third-party vulnerabilities	Vulnerability attacks	DDoS attacks	Tampered websites	Websites infected with Trojans	Overall cybersecurity index
1	Tianjin	0.421	0.227	0.931	1	0.911	0.518	0.883
3	Guangzhou	0.72	0.226	1	0.066	0.598	0.735	0.852
4	Shanghai	0.64	0.08	1	0	0.156	0.8	0.847
5	Beijing	0.625	0.285	0.202	0.116	0.764	0.633	0.831
6	Shenzhen	0.709	0.13	0.606	0.076	0.927	0.668	0.83
7	Chengdu	0.338	0.124	0.991	0.809	0.683	0.51	0.772
18	Chongqing	0.427	0.191	0.271	0.223	0.027	0.465	0.653
25	Urumqi	0.866	0.382	1	0.62	0.471	0.84	0.495
33	Lhasa	0.661	0.143	1	1	0.005	0.701	0.246
35	Haikou	0.45	0.322	1	1	0.103	0.424	0.209

As good and as unique as China's surveillance assets are, the government agencies cannot for now escape its cyber vulnerabilities and the generally uneven state of informatization throughout its large territory. Moreover, they cannot easily escape the historically low per capita ratio between law enforcement officials and citizens, nor citizens' contempt for Chinese police and the legal system. The wide-ranging intelligence collection brief of China's internal security agencies makes them a prime target for attack by adversary intelligence agencies. If a foreign agency can raid processed Chinese intelligence about the internal security situation in the country it need not spend so much effort collecting that information itself. Information in the internal security databases will be of interest to foreign military intelligence and operational planners as well.

To understand the current condition of the cybersecurity of China's PSI system, we have to discuss it under two quite distinct sub-sets: content security (preventing information "pollution") and system security (preventing technical intrusions of networks and computerized systems, and protecting the data). As discussed later, there is reason to believe that the government's heavy emphasis on maintaining content security for political reasons has undermined the necessary attention and resource flow to system security. But as revealed later, China does not do either mission as well as it claims.

Content security has both a technical aspect and social aspect. Sometime around 2001, the Politburo of the CCP agreed that there would not be any purely technical solution in its efforts to censor the internet in China and that there would have to be a highly developed set of social controls as well. This meant that China would pursue technical surveillance and censorship options to the fullest extent possible while developing armies of allies in universities and business enterprises to exercise social discipline to ensure their members did not cross the line. It also meant that in between these two forms would sit battalions of internet police and citizen volunteers supported by new laws and large investigative assets. On the technical side, there are two powerful tools unique to China (Zheng et al. 2013: 2557): a small number of controlled international gateways (only three) and highly efficient filtering capabilities (the "Great Firewall").

However, these is an additional, larger perspective that goes well beyond censorship and social discipline of people who use politically sensitive or banned content. The people who conduct internal surveillance by cyber means also have responsibility to collect internally generated information of high relevance to external security interests to China in its assessment of foreign intelligence topics. This was acknowledged in U.S. congressional hearings in a U.S. specialist's view that "due to its growing internal database, technical sophistication and cyber capabilities, it [MPS] is having a more counterintelligence mission shared with the MSS" (USCC 2016: 6). As Yang (2017: 82) observes, the public security police have a legislated mandate (Article Two of the People's Police Law) to "take the safeguarding of state security as the foremost duty of the police". He concludes that the public security police do "not fall strictly into either the [public] order or political police category" but "simultaneously fight common and political crimes".

The internal cyber-enabled surveillance system is not only directed at people with views that the authorities do not like or at people subscribing to internet gambling or pornography websites, but at the full spectrum of Chinese intelligence interests, especially people who have meaningful ties to targets of foreign intelligence interest. While specialized units exist for cyber intelligence collection outside China, those that conduct cyber "surveillance and monitoring" inside China must be responsive to a wide range of needs in addition to censorship.

The massively large list of intelligence collection priorities for these agencies operating domestically in cyberspace is a very long one and would look something like the list in Box 6.1. The importance of the list is that it is suggestive of a relative low priority for network security of public security systems given the work burdens associated with this list or tasks, adding some corroboration for reports that MPS personnel around China are seriously overworked, and under considerable political pressure on a daily basis.

Box 6.1: Domestic Intelligence Collection Priorities for Cyber Capable Agencies

1. Personal security of the highest level leaders (against possible physical threats, foreign espionage and internal political action—around 50 to 200 high value targets to protect)
2. Security of CCP rule from home threats

   (a) Monitoring labour unrest throughout the work force (thousands of individual targets in critical sectors?)

   (b) Monitoring people who have ties with Taiwanese independence activists in Taiwan and internationally (thousands of individual targets?)

   (c) Monitoring people who have ties with Tibetan independence activists inside China and their links to outsiders (thousands of individual targets?)

   (d) Monitoring people who have ties with Xinjiang independence activists inside China and links to outsiders (thousands of individual targets?)

   (e) Monitoring dissidents and potential dissidents in major mainland cities opposed to Party rule and their links with supporters and like-minded individuals outside China (thousands of individual targets?)

   (f) Monitoring activists and intelligentsia inside China who might organize public ad hoc opposition events in major cities (up to several thousand targets per year?)

   (g) Monitoring workers and community groups involved in ad hoc public protests throughout the country (up to several thousand people per year?)

(h)   Monitoring the police for signs of disloyalty that threaten the leaders or CCP rule

(i)   Monitoring the armed forces for signs of disloyalty that threaten the leaders or CCP rule

(j)   Monitoring Falun Gong activities down to high school level (thousands of people?)

3.   Foreign subversion activities directed into China from outside that are not directly connected to Chinese activists inside the country (hundreds of targets?)

4.   Counter-espionage against foreign spies and their potential agents in China (all diplomats, all journalists, many business people, professionals, teachers, students—up to 20,000 people at any given time)

5.   Security of China from economic crisis, economic disadvantage or economic threat

a.   National economic issues

b.   Trade negotiations

c.   Foreign currency and derivatives positions, stock markets

d.   Plans or foreign investors, both inside and outside China

e.   Counter espionage.

To understand the normalcy of such a large intelligence collection brief, one need only reflect on the Prism program revealed by Edward Snowden and in which NSA was taking a direct feed from telephone companies of calls made in the United States into its computerized databases. Of special note is that in the United States, such action represented by the Prism program was outside the scope of the Constitution and the law, whereas in China, there would be no such obstacles to domestic surveillance of the same kind.

The agencies involved in China's comprehensive public security intelligence (PSI) ecosystem have been identified as comprising two types: intelligence units found in public security comprehensive command centres (CCC) and intelligence units established in the offices of the MPS (Huang 2016: 73). Inside the CCC, there are two main types: the China Crime Information Centre (CICC) and the Super Intelligence System (SIS). Little else is known in the public domain about these centres.

The work horses of the CCP's cybersecurity defence system domestically are "cybersecurity bureaus" under the MPS that are spread around the country. The mission sets of these bureaus are:

1.   supervision, inspection, guidance of information system security work

2.   organization and implementation of security assessments of information systems

3.   investigating and dealing with cyber crime

4.   collations of data on major security incidents

5.   prevention and management of intrusion by viruses and other malware

6.   provision of information system security services and products for management

7. managing information system security training
8. other duties prescribed by laws, regulations and regulations.

These units have been in place for many years. By February 2017, under a policy foreshadowed as early as 2014, they had been augmented by the creation of cybersecurity police units (as mentioned above) in 1116 major firms throughout the country-a move unequalled by any other country and one which reflects a level of governmental insecurity and authoritarian intent seen in few others. According to the newspaper that reported this figure, such a policy "inclines netizens to practice self-censorship to avoid punishment" (Wang 2017). This uniformed force deployed in the private sector provides both technical security support as well as content security support. The latter mission is also supported by an army of "volunteers" throughout large organizations in China who conduct internet patrols since the government sees employers as responsible for ensuring their employees don't breach leadership preferences for content management.

But the lead organization for content security is the Central Propaganda Department (CPD) which organises centralized monitoring and take-down of material deemed inappropriate. The CPD blends with parts of the CAC, the Cyber Emergency Bureau and the Coordination Bureau for Cybersecurity. The Ministry of Industry and Information Technology (MIIT) also has a Cybersecurity Administration Bureau which is responsible for industry supervision in the interests of both content and technical security. It is unclear what role the Ministry of State Security (MSS) has for domestic information security (either technical or content related), but we can assume it has a direct interest in monitoring the possible leaking of state secrets in cyberspace. Since China has a propensity to make even the most anodyne information into state secrets. MSS must be very busy in cyberspace monitoring. Numerous ministries and agencies apart from the MPS, CPD and MIIT play important roles in day-to-day cybersecurity operations. All national ministries and peer agencies have a cybersecurity unit that reports directly to a very senior official but there is little public domain information on how capable these teams are.

As reported by a U.S. specialist, the State Councillor and Minister of Public Security, Meng Jianzhu, announced in 2008 the nation-wide adoption of "public security informatization": the "process of integrating information more closely into police operations, including both domestic intelligence gathering and information management components" (Mattis 2012: 50). At around this time, the Chair of the CPLC, Zhou Yongkang, and Meng's superior, made a tour of some key MPS sites to promote greater attention to the speed of informatization in the public security sector. While there is not a direct or linear relationship between the pace of informatization in an organization and its levels of cybersecurity, the former can—in the absence of other stronger evidence—be taken as a proxy for the latter.

The assessment by a Chinese specialist that the country's PSI system is far from perfect (Huang 2016: 73) seems credible. The author identifies six challenges:

1. the informatization of the intelligence system is inadequate (no unified standard for intelligence reporting and slow updates of intelligence data)

2. inadequate data integration (SIS system covers nearly 10 billion pieces of information, some of which is "disorganized")
3. security risks in the preservation of intelligence data
4. some investigation personnel are incapable of collecting useful information
5. in some places, intelligence collection has not yet attracted enough attention
6. conflicts between the work of public security intelligence and the privacy of citizens.

Another scholar reports three major short comings in their PSI work in general: "clogged intelligence sharing, limited analytical and quality control capabilities, and old ways of policing based on obsolete ideas" (Yang 2017: 91).

While these do not pertain directly to information security of the MPS systems, and they are challenges shared with other countries, these judgements add weight to the view that the deficiencies in MPS information security are likely to be substantial. Two key questions, addressed further below, are: just how large are the information security teams of the MPS offices throughout the country and how well trained are the personnel.

Chinese assessments of the cybersecurity ecosystem inside China's national security agencies reveal high levels of dissatisfaction. In general terms, there was a common view around 2013 that the government sector suffered from the low cybersecurity awareness of its top leaders and employees, particularly on issues like psychology and incentivization (Zeng et al. 2013: 687). This study implied that China's public sector had yet to consolidate the "institutionalization" phase of information security: where the practices became a "natural aspect of the daily activities of all employees" (686).

These judgments were corroborated in a 2017 commentary by an official of the Jintan Branch of Changzhou Public Security Bureau (Li 2017). He described "security vulnerabilities in software", lack of standardization of hardware between MPS offices, resulting potential for data leakage, unsecured nodes and terminals linking to secured systems. He also complained about a lack of cybersecurity awareness among colleagues, access by non-authorized police to MPS systems, improper use of USBs and mobile hard disks, and storage of confidential documents in unsecured computers. The findings of this study reappeared in several later articles attributed to officials of the MPS.

In a 2016 study of public security studies in China, the authors from the Yunnan Police College complained of several shortcomings in public security training and talent building in cyberspace (Jia and Wang 2016: 231–32). These included an over-emphasis on theory and "scant attention to practice" that contributed to a decline in student interest; the rapid pace at which knowledge acquired during studies is overtaken by new technologies, thus making student knowledge outdated; and the related difficulty teachers had in updating their knowledge. The authors lamented the dearth of talents in applied cybersecurity compared with engineering and research.

The authors cited a member of the National Informatization Expert Advisory Committee, Shen Changxiang, who suggests that the government should take a stronger lead in guiding the construction of cybersecurity discipline and accelerate the pace. There was, Shen was cited as saying, a need for innovation in approaches to talent cultivation processes; and significant reform of the pedagogical processes ("build teacher teams to promote education quality"; "improve textbooks and curricula"; and "set up a training platform to strengthen practical exercises for cyber attack and defence") (231).

Other complaints raised include: many defects in the existing intrusion detection of public security defence units (lack of active defence capability and the system's error/leakage rate is high) (Yu and Hu 2013: 185); and cyber monitoring in the western provinces has been relatively backward (Zeng 2010: 18). In fact, the most advanced intelligence sharing system for the MPS in China is only available in 131 cities (mostly those with a population of around one million or greater) out of several hundred cities. Many rural and semi-rural areas will also not have access to this intelligence system.

We can gain further insights from other research papers commenting more positively on future directions but revealing implied or explicit criticisms.

In a short analysis noting that digital data had become the "primary objects of public security intelligence collection", one author reported problems such as enormous workloads, low efficiency, and poor quality (Peng 2017a: 33). Calling out the need for consequential reforms (such as intelligence collection ideas, means, scope, and targets), he foreshadowed other more fundamental measures: "scientifically defining the boundary of public security intelligence collection, widely collecting various types of information, giving full play to people's initiative, comprehensively using artificial [intelligence] collection methods" (36–7). The same author predicted elsewhere that the rising of big data technology would "inevitably exert a profound impact on public security intelligence science and its research paradigm", but that in this process, agencies needed to be able to "minimise negative effects of big data" (Peng 2017b: 82). He called for multidisciplinary research methods to be the science of public security intelligence. He made a heartfelt plea for maintaining a balance between the rush to informatization and more established dispositions: "relying on correlation analysis but ignoring causality analysis, advocating universal data but despising people's intelligence, worshiping technology but neglecting the dominant position of human being" (82). In a third article on a similar theme, the researcher lamented that public security agencies "have not placed sufficient attention to digital data on the Internet" and "their research on cyber technology is developing slowly" (Peng 2017c: 12). In 2015, another researcher had issued a call for training of public security personnel in big data approaches to respond to the opportunities and to improve public security work (Zhang 2015: 11).

An additional source for forming a judgement about the quality of cybersecurity in the MPS is the quality of cybersecurity education possessed by the leaders and mid-level officers in these agencies. This comes from several places, but an essential foundation is what is taught in the public security university, the police universities and the police colleges. At the national level, China has the Public Security University in Beijing, the Police University in Shenyang, the People's Armed Police Force College in Langfang (80 km from Beijing), the Maritime Police College (near Ningbo in Zhejiang province), the Railway Police College, a Correctional Police College, and provincial level colleges in only 16 of the country's 31 administrative regions.

It was only in 2005 that the MoE and the Public Security University set up the first bachelor's degree in public security intelligence (Peng 2017c: 14). In 2011, the Academic Degrees Committee of the State Council and MoE elevated "public security science" to a first level sub-discipline in faculties of law. At that time, public security intelligence science was described as a second level discipline under public security science (12). Only two universities in the MPS system received authorization in 2016 to elevate information security to a Level One discipline: the Public Security University and the National Police University. In 2016 CUAA rankings, only one of these institutions (Public Security University) received a six star ranking in the field of "public security technology". This is in a ranking system of 1–9 stars, the latter number being the highest rank. Only two other MPS-related universities revived a five-star ranking: People's Armed Police Force Academy and the Criminal Police Academy (CUAA 2016). While some students who graduate from such colleges can be excepted to undertake master's degrees, or even Ph.Ds, the numbers of these are probably quite small. In a 2009 article, two researchers said that standards in public security education were much lower than in normal universities and colleges because reforms started later than in normal universities and were building on a weak base and poor conditions (Ma and Zheng 2009: 176).

According to its website, the Public Security University (PSU) is a police academy (a uniformed student body) with high entrance standards. It has 10 colleges, including criminal science and technology college, police information engineering college, and a network security college. It has three teaching departments: a police command and tactical department, a public security information systems, and a traffic management engineering department. It has two research and teaching departments: police combat training and social sciences. The Network Security Defence College in PSU, according to a U.S. press report, is home to 15 labs which test and provide training in a variety of techniques (Gertz 2014). This report suggests, probably incorrectly, that this university trains both PLA and PAP officers in addition to MPS students.

The National Police University (NPU) in Shenyang, according to its website, has ten undergraduate majors, many of which have a cyber element, but one specialising in Cybersecurity and Law Enforcement, which brought together previous courses in pre-existing majors (such as Criminal Investigation and Forensic Science, and Information Security). The university has set up training centres, research centres and laboratories at a variety of levels (ministry, province, and city). It is highly internationalized, trains students from 50 foreign countries, and conducts academic exchanges with similar specialized programs in foreign universities. The staff/student ratio is very impressive: 750 staff (380 full-time teachers, of which 139 are associate professors and 82 are professors), but with around 6000 students at any time, including full-time undergraduates, postgraduates and in-service police trainees. It has 15 departments, of which one is dedicated to investigation of computer crime.

The three named professors in Computer Crime Dept. do not appear to have cutting edge scientific and technical qualifications in information security, but rather specialise in cyber crime investigation and electronic evidence identification (NPU website). But some professors in other departments, like the Public Security Intelligence Department, specialise in subjects like technical reconnaissance. The three named professors in the Politics and Theory Department teach subjects like Marxism, dialectics and Mao Zedong thought, and not subjects on the public policy of cybersecurity.

This brief glimpse into what the two MPS-related information security universities teach and who teaches is a reminder that the capability of Chinese cyber defence forces depends on a huge diversity of scientific and academic specializations. Another good reminder of this can be found in a literature review of Chinese and international studies on the subject which lists 34 distinct subfields under five broad headings (cryptography, network security, information system security, information content security, and information confrontation (Zhang et al. 2015: 4–5). To reflect such diversity of the subject of cybersecurity, a 2016 Chinese study proposed a four by four matrix (16 elements in total, all overlain by four theoretical approaches involving new technologies: Cloud computing; Internet of things; quantum cryptography; and post-quantum cryptography) (Luo et al. 2016: 941).

An illustration of how slowly the Chinese education system can react to this diversity can be found in the list below (a very short one) of newly approved university and college courses in China in 2014 in the field of public security (MOE 2017):

- Two years of public security audio visual technology in the Public Security University
- four years of foreign police law in the China Criminal Police University
- two years of network security and law enforcement engineering in the China Criminal Police University
- four years of information security engineering four years in the Public security police college
- four years of network security and law enforcement (at an unspecified location)

Given the cyber power declaration of Xi in February 2014, one might have expected this list to be longer and to reflect the vast diversity of the field more comprehensively.

The quality of information security education in the MPS system has been criticized by specialists on a number of grounds, including inadequate practical training, uninteresting content, poor technical facilities, and the fact that students are often called away from studies for MPS or police operations (Leng and Zhu 2015: 72).

As mentioned in Chap. 2, university education in cybersecurity is complemented by other formats (vocational colleges, continuing education colleges, employer-provided training, and private training centres). An author from an MPS college in Wuhan (Jiang 2016: 75) gives some insight into the evolution of continuing education in cybersecurity for public security officers:

2005: 3rd Research Institute of MPS set up Shanghai Haidun Security Technology Training Centre under the National Anti-Computer Intrusion and Anti-Virus Research Centre

2011: 3rd Research Institute set up a cybersecurity branch of the National Public Security Education and Training Network Online College to provide training for the police throughout China

2013: MPS assigned the National Research Center of Anti-Computer-Intrusion and Anti-Computer-Virus Research Centre the title of the ministry's national level continuing education base of technical professionals. The continuing education base was the first one in the information security field, and it is the only one for the MPS
2014: The MPS set up a program for Internet police in the centre
2016: Student completions in the training base in the areas of cybersecurity awareness in the previous two to three years exceeded more than 100,000 people; while the numbers for professional certifications in information security exceeded 9000. Some 4000 people were trained as information security evaluators and another 4000 received technical knowledge update training.

The mention of the 3rd Research Institute above in organizing vocational training is one indicator of the increasingly important operational role this organization plays within MPS and national cybersecurity implementation. The Institute is a power house of capability inside MPS and engages in a range of activities not normally associated with such research centres in other countries, such as the certification of security levels of enterprises. Each year since 2011, the institute has co-hosted, with the Cybersecurity Defense Bureau of the MPS, a national level Conference on the Construction of Information Security Level Protection Evaluation System. Beginning in 2016, it co-hosted a national level annual "Conference on China's Information Security Services". For this, its partners included the National Quality Supervision and Testing Centre of Security Product for Cyber and Information Systems. It hosts other ad hoc conferences working closely with a variety of stakeholders, such as the Shanghai Information Network Security Administration Association and the Shanghai Research Centre for Disaster Prevention and Security Strategy, for a 2016 conference on Counter Measures for Cybersecurity Risk. In 2016, it partnered with the Cybersecurity Coordination Bureau of the Office of the Central Leading Group for Cyberspace Affairs, the Cybersecurity Administration of the Ministry of Industry and Information, the State Encryption Administration, the State Secrets Bureau, and the Chinese Academy of Sciences in the Fifth National Information Security Level Protection Technology Conference, an annual series beginning in 2012 in which it has not always been involved.

The 3rd Research Institute of the MPS has about up to 2000 staff, of whom 1600 are scientific research personnel. Apart from its Information Security Technology Department and a National Information Security Division, it hosts the National Anti-Computer Intrusion and Anti-Virus Research Center, the MPS Information and Network Security Key Laboratory, Network Events Early Warning, Prevention and Control Technology Laboratory, and an electronic data forensic laboratory set up in 2008 (the first of its kind in China). But the mission of the institute covers some fields, such as industrial alarm systems, movement detection and counter-terrorism technologies, that have little to do with information security. According to the CNKI database, the Institute researchers have not been very active, with no more than one article per doctorate qualification but the Institute website claims 140 publications "in recent years". In September 2017, it announced a recruitment drive for 40 graduates (Ph D preferred) to enter in 2018.

The institute's highest priority appears to be developing technologies that track and identify Chinese citizens, vehicles, and products and analysing those movements for intelligence purposes. Its work in information security policy and standards development may be its biggest contribution to the country's cybersecurity.

6.3   Critical National Infrastructure

In terms of open-source studies, one of the earliest serious interventions in China's public policy for critical information infrastructure (CII) protection was by the Information Security Law Research Centre (ISL) of Xian Jiatong University when it published a "blue paper" on the subject in 2012 (XJTU 2012). As noted in a 2014 study co-written by the present author, the report "does not give much insight into policy" but "demonstrates the relatively recent focus by China" on this subject (Austin et al. 2014: 14). The blue paper was presented as a quick guide, an "introductory note for the international community to understand China's laws, regulations and policies for the protection of critical information infrastructure."

The paper notes that "Concepts like 'critical infrastructure' (CI) and 'critical information infrastructure' can be found nowhere in China's laws and regulations" (1). At the same time, it documents a number of laws and government decisions that relate to the goal of CII protection, including the Emergency Response Law of 2007 that is not specifically dedicated to information assets. Apart from mentioning the 1994 law on a national grading system for information security (MLPS referred to above) and various international commitments by China on information security in the framework of the Shanghai Cooperation Organization (SCO), the Blue Paper takes as its major reference point a set of 18 standards issued in 2010 by China's Information Security Standardization Technical Committee addressing Public Key Infrastructure Security, Cryptography and Certificate Authentication Systems, and a Guide for Graded Protection of Information Systems (XJTU 2012: 4).

The blue paper identified the following priority sectors:

- government affairs information systems
- CCP affairs information systems
- livelihood sectors (finance, banking, taxation, customs, auditing, industry, commerce, social welfare, energy, communication and transportation, and national defence industry)
- educational and governmental research institutes
- public communications, such as radio and television.

The blue paper singled out a number of agencies who should lead in CIIP (XJTU 2012: 5) as listed below, and provides some detail on the roles of the first five in this list:

- Ministry of Public Security
- State Cryptography Administration

- National Administration for the Protection of State Secrets
- General Administration of Quality Supervision, Inspection and Quarantine
- Ministry of Commerce
- People's Bank of China
- China's Securities Regulatory Commission (CSRC)
- China's Banking Regulatory Commission (CBRC)
- China's Insurance Regulatory Commission (CIRC)
- National Development and Reform Commission
- Ministry of Science and Technology
- Ministry of Industry and Information Technology

This list appears to assign unusual prominence to the financial services sector.

The above list does not address responsibility for operational coordination, and this is covered subsequently in the blue paper, with CNCERT/CC in the lead for early warning (working with relevant security agencies) (XJTU 2012: 10–11). Supporting roles are played by the Information Security Certification Centre, Information Technology Security Evaluation Centre, the Quality Supervision and Test Centre of Security Product for Computer Information Systems (in MPS), the Anti-Virus Products Testing and Certification Centre, and the National Research Centre for Anti-Computer Invasion and Virus Prevention (XJTU 2012: 11–14).

The working group which prepared the blue paper included representatives from the Protection Bureau of the Ministry of Public Security, the First and Third Research Institutes of the Ministry of Public Security, leading private sector corporations (including Microsoft, Intel, Qihoo and Huawei), government and CCP agencies, and researchers. In the blue paper, the research centre said it was the "executive body for China's Cloud Computing Security Policy and Law Working Group." The centre is the main organizer for China's Information Security Law Conference (held each year since 2010) and runs China's Information Security Law website.

The escalating leadership interest in the national security aspects of cyberspace after the blue paper was issued in 2012, especially the focus brought on by the Snowden revelations beginning in mid-2013, has brought new attention to CII protection in China. Yet its policy community has not moved as rapidly as its equivalent in the United States. This slow-mover trend has been reflected in the research community as well. According to a 2015 study on research into industrial control systems (ICS) as an important part of national critical infrastructure, domestic research on security in this field "has just begun" (Fan et al. 2015: 1) One of the main recommendations from the authors was developments of a "common ICS security audit system" (6), but this was one of a long list of action items based on international practices, including defense in depth, active defense measures, and passive defense measures. The authors repeated the view mentioned earlier that there is inadequate research on the costs of cybersecurity of the sort needed to inform decisions on risk management (5). A British organization reported in 2015 that China's 1994 law on MLPS functioned as China's policy in this area; and that "there is no publicly available legislation or policy in place in China that includes a clear definition for "critical infrastructure protection" (BSA 2015). The same source noted that China lacked any private-public mechanisms of the sort regarded as essential for CII protection in many other countries.

In June 2015, the government claimed full responsibility for CII protection in the first draft of the new cybersecurity law, and set out for the first time a broad definition of CII:

- basic information networks providing services (such as public correspondence and radio and television broadcast)
- critical information systems for important industries (such as energy, transportation, water use, and finance)
- public utilities and essential services (such as electricity, water and gas, medical and sanitation services, and social security)
- military networks and government affairs networks at city level and above
- networks and systems owned or managed by network service providers with massive numbers of users.

As USITO has noted, a Draft on Civil Aviation regulations issued in February 2017 seems to have accepted that civil aviation network information systems are CII assets, since the draft reflects central planks of the cybersecurity law such as the MLPS (Articles 10 &11), the "secure and controllable" principle (Article 14), data localization Article 28), and passenger information protection (Articles 35–37) (USITO 2017).

A key attack trend identified in 2016 by CNCERT/CC was an increase in attacks on industrial control systems. The most likely perpetrators of the majority of such attacks are foreign governments or their proxies. By December 2016, the agency had recorded 1036 vulnerabilities in such systems, of which 173 had been identified in 2016 (CNCERT/CC 2016). Of special note to Chinese perceptions of unwanted dependence on foreign technologies, the agency counted 2504 unique control systems networked outside the country, involving suppliers like Siemens, Rockwell, Schneider, Omron, HP, Advancetech and Vykon Controls. The total number of individual SCADA units in this count was 880,000, linked with 1610 unique IP addresses in 60 countries. Protocols involved included S7 Comm, Modbus, SNMP, Ethernet, Fox, and FINS. On the data available, it seems that the dominant supplier, Siemens, with 44% share of these units, was responsible for 14% of the vulnerabilities found. More than 50% of the vulnerabilities found were outside the top five suppliers (all of which were non-Chinese). The presumption we can draw from this is that the top-line foreign suppliers of SCADA are on average more cyber secure than domestic suppliers.

In March 2017, in the light of record DDoS attacks in 2016 (in excess of 1 terabyte per second), one of China's leading private cybersecurity companies, NSFocus, issued a study of China's exposure to vulnerabilities in the Internet of Things (IoT), the main new purveyor of such attacks (NSFocus 2017). It found that among 12 device types, China had around 25% of the world's exposed routers (the main source of exposure globally)—four million out 16 million (4). The man exposure came from domestic brands (3). It also found that China had 16% of exposed web cameras, and almost 100% of exposed network bridges. Its most important conclusion may be that

"The reason that most devices are probed is that they are deployed on the Internet with no changes to default settings" (3), a fact that reveals low security awareness on the part of the vendors and consumers.

By April 2017, China had held its second Critical Information Infrastructure Level of Protection Seminar (China News 2017) to study international experience and debate Chinese planning. The seminar was co-organized by the Ministry of Public Security, supported by Office of the Central Leading Group for Cyberspace Affairs, the State Secrecy Bureau, State Cryptography Administration, the Beijing Public Security Bureau, and Qihoo 360 Enterprise Security Group, as the domestic cybersecurity industry leader. The Qihoo involvement seems to have centred on research and recommendations, developed jointly with MPS officials, to enhance situational awareness, particularly the role of advanced technologies in monitoring, reporting and warning. 360 reported that it was working to assist Public Security offices, large industries and large enterprises to establish a number of situational awareness systems.

In late June 2017, in response to a requirement mandated by the cybersecurity law, the Office of the Central Leading Group for Cyberspace Affairs released an emergency response plan for security incidents that was central to government strategy for CII protection but not limited to that. Its purpose was to "improve handling of cybersecurity incidents, prevent and reduce damage, protect the public interest and safeguard national security, public safety and social order" (China Daily 2017). It set out four levels of security alert and response systems from 'general' to 'extremely serious'. The latter included incidents that "paralyse many important internet and information systems". The announcement said that serious incidents would "trigger measures including the establishment of an emergency headquarters, 24-h monitoring and multi-departmental coordination".

On 10 July 2017, when the CAC issued the Draft Regulations on CII Protection to implement the provisions of the law, in Articles 21 and 22, it shifted the responsibility fully to individual operators (whether governmental, state-owned or private sector). The draft stipulates that "when CII is constructed, it must have security technology measures… that are planned, constructed and used" (Article 21); and the "main responsible persons are the first responsible person for CII security protection within a work unit … and they are completely responsible for CII security protection work within their work units" (Article 22) (CAC 2017). Chapter 7 of the draft provides legal penalties of a civil kind for failings, including use of uncertified or uninspected assets. While these penalties are not overly harsh in monetary terms by Western punitive standards (less than one million RMB), they do include provision for a lifetime ban on "working in critical information infrastructure security management and network operations key positions" where a criminal offence is involved. Article 50 of the draft imposes a draconian standard making supervisors and managers liable for their employees where the latter are found to have obtained or accepted bribes, to be responsible for "neglect of duty, abuse of authority; unauthorized disclosure of information or data files of key information infrastructure"; or "other acts that violate statutory duties".

The draft regulations slightly modified the definition in the law to include the activity to be monitored ("CII whenever its destruction, ceasing to function or leakage of data may gravely harm national security, the national economy, the people's livelihood and the public interest"). They also indicated the field of coverage to include "work units providing cloud computing, big data and other such large-scale public information network services"; "research and production work units in sectors and areas such as national defense science and industry, large-scale equipment, chemistry, food, drugs, etc." and "other news outlets akin to radio and television". In September 2017, a senior official of the MPS announced that China is considering the adoption of a "cybersecurity classification protection regulation" for CII infrastructure protection (Cao 2017).

6.4 Leadership: Internet Terror

The limits of cyber safeguards inside China's intelligence and security agencies, including those charged with protecting the leaders, have been spectacularly revealed to the world in separate incidents every year between 2012 and 2016.

In 2012, at a time of leadership transition in China, two news stories in U.S. news outlets in quick succession drew attention to the wealth of the immediate family of incoming President Xi Jinping and outgoing Premier Wen Jiabao (Austin 2014). Though these news reports were assembled largely from public record documents and human sources, it is highly likely that some details, especially those of pseudonyms of family members and their business holdings under those names, had been obtained by cyber espionage of some kind. In February 2013, a U.S. company, Mandiant, operating on a brief from the U.S. government, released a detailed report on the operations of a PLA intelligence unit that exposed huge weaknesses both in its system cybersecurity and the personal cybersecurity of several of the unit's staff (Mandiant 2013). Less than four months later, the Snowden leaks mentioned above rocked leadership confidence in their cybersecurity. In 2014, the International Consortium of Investigative Journalists (ICIJ 2014) identified 22,000 separate clients residing in China (including Hong Kong) who held offshore accounts in the tax havens of the British Virgin Islands after a dump of cyber data, presumably by a foreign intelligence agency trying to expose tax evaders. In 2015, the U.S. legal system brought indictments against five PLA personnel previously identified in the Mandiant report but revealed important new information on the detail of the cyber espionage, further demonstrating massive basic deficiencies in security of the cyber units involved (United States 2015). In 2016, the leak of data from Panamanian law firm, Mossack Fonseca, revealed details on 40,000 companies, past and present, from China and Hong Kong, some with ties to close family members of current and former leaders of China (Garside and Pegg 2016).

This pattern of revelations bears out the view of Chinese leaders that "cyberspace has turned into the strategic cornerstone of national security in the Internet age"…. "new situations, new problems, new features and new trends are expressed in most cases" (Cai 2015: 477).

The term "internet terror" (*wǎngluò kǒngjùzhèng*) has been used in newspapers in China to describe the practice of using leaked information to affect political careers and personal lives, though more in the context of citizens' use of the internet to expose corruption by mid-level officials, including work colleagues (Bandurski 2010). But the leaders now know that it affects them, and their hold on power, as well. The difference is that the primary actors they fear are foreign or Chinese dissidents abroad. The leaders also recognise the near certainty that there is a Chinese Edward Snowden out there who will deliver an even greater information catastrophe to them than Snowden did to the NSA. They also fear that one day soon, the U.S. intelligence community, with its massive cyber surveillance capability, will link up with investigative journalists or other activists to publish sensitive information about the leaders on such a scale that the Community Party itself will be discredited almost overnight.

The leaders have images from 1989 in their minds: the Tiananmen protests and the collapse of Communist Parties in Eastern Europe, including the murder of deposed Romanian leader Nicolae Ceaucescu. Now they fear the next wave of resistance will occur online. Indeed, the U.S. government in 2010 offered funding for Falun Gong Internet activity against the Chinese government. There has even been a question of the loyalty of China's cyber intelligence services to the highest level of the regime running through 2012 and 2013. A political scandal erupted around Zhou Yongkang, a former member of the Standing Committee of the Politburo, who had only stepped down in November 2012 from that post and his role as Chair of the Central Political and Legal Commission (CPLC). In effect, Zhou's role and power was equivalent to the combined roles and power in the United States of Director of Central Intelligence, Director of National Intelligence, Director of the FBI, and Attorney General. The CPLC is a CCP body, similar to its military cousin, the Central Military Commission, and it remains arguably the single most powerful institution in the entire country. It controls all police, all courts, all judges, and all legislation, as well as non-military espionage for internal and international security. It controls all encryption and technical standards for China's governmental apparatus. Zhou was eventually convicted of corruption charges in 2015 and jailed, but the evidence of his abuse of power by cyber espionage against his fellow leaders had emerged by 2012, and was confirmed through the investigation of the former police chief in Chongqing, Wang Liqun, the same year. China's leaders had never been subjected to cyber espionage from their own peers as they had been in 2012 after stepping up their investigations of Chongqing leader and Party chief, Bo Xilai, a close ally of Zhou. Bo was already a Politburo member aiming for elevation to the Standing Committee. So, internal control of cyber espionage assets became the highest priority of the new Party leadership installed at the Party Congress in November 2012.

Behind the scenes in China, the leaders have moved aggressively to shore up cybersecurity arrangements affecting their personal lives. But all indications are that this is an exercise doomed to failure. There is now no single issue more sensitive in China than internet reporting on the leaders. To counteract that imminent threat, China's leadership has tried the route of technical surveillance by any means and of anyone.

These considerations give rise to a possible process of action and reaction. This mix of insecurity and conjecture could possibly lead to an information war. The Chinese Foreign Ministry has already called into question the motives of the ICIJ, meaning a presumption that the organization is trying to dismantle Party legitimacy in China. As the terabyte leaks affecting China's political class accumulate, the leaders' insecurity will also increase. One thing is for sure: Chinese leaders believe the international information contest is moving to new levels. Issues of ethics and legitimacy long considered settled by the Chinese leadership are now at risk in novel ways either because of the very large scale of leaks themselves or the scale they can take on through new Internet-based media.

6.5 Conclusion

The state of government cyber security in China is in broad terms weak to very weak. This is an inevitable reflection of the country's late start in informatization and its lack of attention to the cyber skills deficit until the most recent times. It is also as likely caused by an over-emphasis on content security to the detriment of system security.

At the same time, we can assume that some areas of governmental cyber security are much stronger than others. For example, since the Chinese leaders value above all else their own security, we can imagine that network systems and databases that support leadership communications and operations are much more secure than others. The same might be said of strategic military communications. Yet we have little reliable evidence to go on. We can repeat the observation that Unit 61398 of the PLA charged with cyber espionage had very low standards of cyber security in 2013. We can also note open source reports from Chinese analysts that cybersecurity inside the Ministry of Public Security is in the broad quite weak. The Chinese leaders have recognized the severe shortcoming in governmental cybersecurity but given the scale of the problem (shared with other governments throughout the world) it will take them a decade or two to begin to approach high standards in most sectors of government operations. It is weaknesses in this area of government cybersecurity capability that will be one of the most powerful drivers of China's willingness to continue to rely on foreign vendors, including American corporations, such as Microsoft, Cisco and IBM.

References

Austin G (2014) Terabyte leaks and political legitimacy in the U.S. and China. The Globalist, 24 Jan 2014. https://www.theglobalist.com/terabyte-leaks-political-legitimacy-u-s-china/

Austin G, Cappon E, McConnell B, Kostyuk N (2014) A measure of restraint in cyberspace: reducing risk to civilian nuclear assets. EastWest Institute, New York. https://www.eastwest.ngo/sites/default/files/A%20Measure%20of%20Restraint%20in%20Cyberspace.pdf

Bandurski D (2010) Internet terrors. China Media Project, 29 Oct 2010. http://chinamediaproject.org/2010/10/29/internet-terrors/

BSA (2015) China. Asia-Pacific cybersecurity dashboard. The Software Alliance. http://cybersecurity.bsa.org/2015/apac/assets/PDFs/country_reports/cs_china.pdf

CAC (2017) Critical information infrastructure security protection regulations, Cyberspace administration of China. China copyright and media. (trans: Webster G, Triolo P, Creemers R). Updated 2 July 2017. https://chinacopyrightandmedia.wordpress.com/2017/07/10/critical-information-infrastructure-security-protection-regulations/

Cai C (2015) Cybersecurity in Chinese context: changing concepts, vital interests and cooperative willingness. China Q Int Strateg Stud 1(3): 471–496 (October 2015)

Cao S (2017) China to mobilize net users for cyber security. Global Times, 13 Sept 2017. http://www.globaltimes.cn/content/1066130.shtml

Ceprei Laboratory (2017) Big data city network security index report. 360 cyber security center. http://zt.360.cn/dl.php?filename=%E5%A4%A7%E6%95%B0%E6%8D%AE%E5%9F%8E%E5%B8%82%E7%BD%91%E7%BB%9C%E5%AE%89%E5%85%A8%E6%8C%87%E6%95%B0%E6%8A%A5%E5%91%8A.pdf

China Daily (2017) Nation rolls out emergency cybersecurity plan, 29 June 2017. http://usa.chinadaily.com.cn/epaper/2017-06/29/content_29932460.htm

China News (2017) Critical information infrastructure level protection seminar held. 360 situational awareness achievements recognized. http://www.chinanews.com/it/2017/04-26/8209552.shtml

CNCERT/CC (2016) A summary of China's internet security situation in 2015. National Computer Network Emergency Technology Processing Coordination Center, April 2016. In Chinese http://www.cac.gov.cn/files/pdf/wlaq/Annual%20Report/CNCERT2015.pdf

CUAA (2016) Public security technology rankings list of China's universities in 2016. http://www.cuaa.net/cur/zhuanti/news.jsp?information_id=130796 (In Chinese)

Fan X, Fan K, Wang Y, Zhou R (2015) Overview of cyber-security of industrial control system. In: 2015 international conference on cyber security of smart cities, industrial control system and communications (SSIC). 1–7http://toc.proceedings.com/27630webtoc.pdf

Garside J, Pegg D (2016) Panama papers reveal offshore secrets of China's red nobility. The Guardian, 6 April 2016. https://www.theguardian.com/news/2016/apr/06/panama-papers-reveal-offshore-secrets-china-red-nobility-big-business

Gertz B (2014) Chinese police university trains Beijing hackers. Free Beacon, 20 March 2014. http://freebeacon.com/national-security/chinese-police-university-trains-beijing-hackers/

Huang S (2016) Research on the problems and countermeasures of public security information application. J Hubei Police Acad (2):71–75 (In Chinese)

ICIJ (2014) Leaked records reveal offshore holdings of China's elite. The Panama Papers, International Commission of Investigative Journalists. https://www.icij.org/investigations/offshore/leaked-records-reveal-offshore-holdings-of-chinas-elite/#

Jia X, Wang J (2016) Discussion on the reform of public security information security personnel training mode based on the teaching of network attack and defense. Information and Computers 9:231–233 In Chinese

Jiang X (2016) Research and application of public security continuing education in information security. J Wuhan Public Secur Cadres Coll 30(1) (In Chinese)

Leng J, Zhu B (2015) Professional practice model reforming of information security specialty in public security college. In: International conference on mechatronics, electronic, industrial and control engineering (MEIC 2015). Atlantis Press, pp. 71–74

Li H (2017) Discussion on public security computer network and information security. In: Network security technology and application magazine. 2017(8) In Chinese. http://www.xueshu.com/wlaqjsyyy/201708/29810655.html

Luo Y (2011) Study on the current situation of information security and counter measures in China. Energy Procedia 5: 392–396. https://www.sciencedirect.com/science/article/pii/S1876610211010034

Luo J, Yang M, Ling Z (2016) Network space security system and key technologies. Sci China: Inf Sci 46(8):939–968 (In Chinese)

Ma D, Zheng L (2009) Research into the status quo of learning strategies of college students and blended learning strategy. In: Hybrid Learning and education, pp. 175–185 (In Chinese)

Mandiant (2013) APT1: exposing one of China's cyber espionage units. https://www.fireeye.com/content/dam/fireeye-www/services/pdfs/mandiant-apt1-report.pdf

Mattis P (2012) The Analytic Challenge of Understanding Chinese Intelligence Services. In" Studies in Intelligence. 56(3):47–57. https://www.cia.gov/library/center-for-the-study-of-intelligence/csi-publications/csi-studies/studies/vol.-56-no.-3/pdfs/Mattis-Understanding%20Chinese%20Intel.pdf

MoE (2017) 2014. Undergraduate record or approval results of ordinary colleges and universities, 25 Feb 2017. http://www.jwz86.com/en/index.php?m=content&c=index&a=show&catid=6&id=34

NSFocus (2017) An analysis of exposed IoT technologies in China. NSFocus technologies. http://blog.nsfocusglobal.com/wp-content/uploads/2017/05/ExposedIoT-AssetsChinaMarch-2017.pdf

Peng Z (2017a) On public security intelligence collection based on big data. In: Research on library science, pp. 33–48. http://en.cnki.com.cn/Article_en/CJFDTOTAL-TSSS201709007.htm (In Chinese)

Peng Z (2017b) On the research paradigm of public security intelligence science under big data environment. Libr Inf 1:82–86. http://en.cnki.com.cn/Article_en/CJFDTOTAL-BOOK201701011.htm (In Chinese)

Peng Z (2017c) On the study object of public security intelligence science. J Intel 36(4):12–17. http://en.cnki.com.cn/Article_en/CJFDTOTAL-QBZZ201704003.htm (In Chinese)

USITO (2017) CAAC drafted new security measures in line with CSL, 20 Feb 2017. http://www.usito.org/news/caac-drafted-new-secuirty-measures-line-csl

United States v. Wang Dong et al United States District Court, Western District of Pennsylvania, filed 1 May 2014 under seal, Criminal No. 14–118. http://www.justice.gov/iso/opa/resources/5122014519132358461949.pdf

United States (2015) District Court Western District of Pennsylvania. United States of America v. Wang Dong et al. Grand Jury Indictment. 1 May 2014. https://www.documentcloud.org/documents/3214423-2014-05-01-USA-v-Wang-Dong-Et-Al-Indictment.html

USCC (2016) China's intelligence services and espionage operations hearing before the U.S.-China economic and security review commission, 9 June 2016. https://www.uscc.gov/sites/default/files/transcripts/June%2009%2C%202016%20Hearing%20Transcript.pdf

Wang H (2017) 1116 cyber-security police units have been set up nationwide. Beijing: Guangming 14 Feb 2017. http://epaper.gmw.cn/gmrb/html/2017-02/14/nw.D110000gmrb_20170214_2-04.htm (In Chinese)

Wong KC (1994) Public security reform in China in the 1990s. In: Brosseau M, Lo CK (eds) China annual review 1994. Chinese University Press, Hong Kong, pp. 5.1–5.33

XJTU (2012) China's protection for critical information infrastructure. Blue paper. Xian Jiaotong University Information Security Law Research Centrehttp://www.infseclaw.net/news/html/817.html

Yang Z (2017) China's public security intelligence: progress, challenges, and prospects. Georgetown J Asian Aff. https://asianstudies.georgetown.edu/sites/asianstudies/files/documents/gjaa_3.2_yang.pdf (Spring 2017)

Yu Y, Hu W (2013) Intrusion detection algorithm design in the public security system for network monitoring. Bull Sci Technol 29(6):185–190 (In Chinese)

Zeng X (2010) Research on management of internet by police in west of China. Netw Secur Technol Appl 4:18–20 (In Chinese)

Zeng Z, Yang K, Zhang Y and Zhou P (2013) Increasing Employees' Awareness and Enhanc-
  ing Motivation in E-Government Security Behavior Management. In: 2013 Fourth International
  Conference on Digital Manufacturing & Automation. 684–687
Zhang H, Han W, Lai X, Lin D, Ma J, Li J (2015) Survey on cyberspace security. Sci China Inf Sci
  58(11):1–43 (In Chinese)
Zhang L (2015) The paradigm change of public security intelligence studies from the per-
  spective of "big data." J Intell 34(7):9–28. http://en.cnki.com.cn/Article_en/CJFDTOTAL-
  QBZZ201507003.htm (In Chinese)
Zheng A, Song P, Han B, Zheng M (2013) Reflection of the nation cybersecurity's evolution. In:
  Applied Mechanics and Materials, pp. 347–350 (In Chinese)

Chapter 7
Grading National Cybersecurity

Abstract There are many useful approaches to evaluating different aspects of a country's ability to provide for its security in cyberspace. Some focus on more technical aspects, while others include broader social and political issues. This chapter documents several assessments of the state of cybersecurity in China, looking first at Chinese assessments and then at international points of view. The chapter concludes with a discussion of China's performance against the nine strategic tasks identified by the December 2016 Cybersecurity Strategy and canvassed in Chap. 1.

7.1 Qualitative Approaches Preferred

Basic statistics and important policy analysis can be found in Chinese official sources, some of which are available in English. These include the China Computer Emergency Response Team (CNCERT), which publishes annual, monthly and weekly reports, and the China Internet Network Information Centre (CNNIC), which has been publishing biannual reports and weekly reports since 1998. In addition, there are annual year books on China's information security (1999 to 2016) produced by the Information Technology Industry Association of China, "blue books" on the development of cyberspace security in China (two issues: 2015 and 2016), and blue books on global information security (three issues: 2014–15, 2015–16, and 2016–17). These works, referenced throughout the book, represent a variety of official and academic views. More recently, the private sector in China has taken up the challenge of producing its own assessments of cybersecurity in the country, and several of these reports (Aqniu, Qihoo 360, Antiy) have been referenced in earlier chapters.

As early as 2000, according to a reliable source, China had set in place a grading system to evaluate its information security, that year assessing its performance as somewhat middling: on a national index in which the highest value was 9, the country was assessed at 5.5, or "somewhere between relatively secure and slightly insecure" (Qu 2010). There appears to be no other public information on this index and one Chinese source reports that the government does not use it any longer. In the meantime, much has changed to affect how China views the security of the global information ecosystem and its impacts on its own security.

© The Author(s) 2018

G. Austin, *Cybersecurity in China*, SpringerBriefs in Cybersecurity,
https://doi.org/10.1007/978-3-319-68436-9_7

As affirmed in the Preface, cybersecurity is a socio-technical phenomenon that is multi-dimensional, multi-stakeholder, highly dynamic, and differs for a number of specific mission sets as diverse as enterprise security and fighting cyber-enabled warfare. There is no single assessment system yet devised that captures the complexity of such a phenomenon.

Austin and Slay (2016) have proposed a checklist for evaluating national cybersecurity policy that is reproduced in Box 7.1. It highlights the need for mature and consistent scholarly analysis of all elements of the complex, multi-layered montage and depends on an implied multi-disciplinarity. It also calls for participation by key stakeholders in articulation of the threats they see and how they respond. Such an assessment on the scale and scope advocated by Slay and Austin has not yet been undertaken in one place for any single country in the world. A key missing link in most countries is evaluation of the performance in cyber security of key stakeholder groups.

Box 7.1: The Checklist

1. Consistent articulation of the different domains of cybersecurity (crime, harassment and bullying, critical infrastructure resilience, espionage, warfare); of the many dimensions of cybersecurity (technical, human, social and legal); and how different sections of the society must bear differentiated responsibilities.
2. Consistent and comprehensive articulation of the threat environment and variegated response options.
3. A comprehensive suite of governmental, cross-sector, private-public, professional and civic organizations active in cybersecurity.
4. National consensus on where to draw the line between sovereign capabilities and the global communities of practice (including R&D).
5. Effective monitoring of business and economic threats and rapid response capabilities at the enterprise level, including large corporations and SMEs.
6. Nation-wide preparedness for the unlikely but credible threat of an extreme cyber emergency affecting the civil economy or national security interests (including international aspects).
7. Effective response capabilities for social threats (crimes) against individuals, including children and other vulnerable groups.

There is no strong consensus globally or in China on the best ways of assessing a country's security in cyberspace. On the one hand, an analyst is left to find whatever combination of qualitative and quantitative assessments that may be credible from inside or outside the country. On the other hand, the analyst can also strive to round out, augment or amend existing approaches to better describe the problem. This chapter follows both avenues, but is informed by the aspirational benchmarking suggested by Austin and Slay.

7.2   Joint China/ITU Assessments

The Chinese government has been working for a number of years in partnership with the International Telecommunications Union (ITU) in the latter's development of an assessment framework of cybersecurity based on the commitment of governments to appropriate policies and to "cyber wellness". It is intended as a mechanism to help countries improve their performance in this area of policy. In 2017, the ITU issued its second study of international comparisons in this series (ITU 2017). The Global Cybersecurity Index (GCI) developed by the ITU has five pillars that emerged from the ITU's first global consultation on cybersecurity, a High Level Expert Group captured in the Chair's Report, *Global Cybersecurity Agenda* (ITU 2007). The Chairman issued a report because the group could not reach consensus. The five pillars are: legal, technical, organizational, capacity building, and international cooperation. The buy-in of the Chinese government to this ITU process is evidenced in part by the fact that a Chinese national, Zhao Houlin, is the current Director General of the ITU, an indicator that China sees the organization and its activities and pronouncements as amenable to its guidance. The first GCI study published in full in 2015 was based on consultations in 2013–14 (ITU 2015). It published its initial results with the caveat that they have "a low level of granularity since it aims at capturing the cybersecurity preparedness of country and NOT its detailed vulnerabilities" (ITU 2014). China was ranked equal 14th, but after 46 other countries. The report noted (134–5) that "China does not currently have any national governance roadmap for cybersecurity", "any officially recognized national or sector specific certification body for cybersecurity", or "an officially recognized national or sector-specific benchmarking exercise or body". The 2017 study involved an expanded set of indicators and much wider consultation, including a survey of 134 participating national governments.

In this study, China was assessed to sit outside the group of 21 "leading" countries, which had GCI scores above the 90th percentile because of their demonstrated high commitment to all five pillars of the index (ITU 2017: 13). China was assessed to be in the "maturing stage", in a group along with another 76 countries with GCI scores between the 50th and 89th percentiles: They were assessed to "have developed complex commitments, and [to] engage in cybersecurity programmes and initiatives". China was ranked 32 in the world, and below countries like Qatar, Mauritius and Mexico (ITU 2017: 51–8). China's score was 0.624 compared with the first-tanked country Singapore with 0.925, and the second ranked country, the United States with 0.919 (ITU 2017: 60–1).

The GCI offered assessments of each country under the five pillars in terms of red, amber or green. China was assessed at green in two of the five pillars (technical measures and domestic capacity building) and amber in three (legal measures, organizational measures, and cooperation). To reach the overall assessment under each pillar, the GCI relied on review of sub-indicators under each pillar (25 sub-indicators in total). China's results for the sub-pillars can be seen in Table 7.1. With reference to the mention of national cybersecurity strategy in the red column below, China

Table 7.1   China's GCI 2017 assessments by sub-indicator of the ITU CGI

Green	Amber	Red
Cybersecurity legislation	Cyber crime legislation	Cybersecurity strategy
Training	Standards of organizations	Cybersecurity metrics
National CERT	Standards of professionals	Multilateral agreements
Government CERTS	Professional training courses	Public-private partnerships
Sectoral CERTS	Education programs	
Child protection online	Bilateral agreements	
Responsible agency		
Standards organizations		
Cybersecurity good practices		
R&D programs		
Public awareness campaigns		
Incentive mechanisms		
Home-grown industries		
International participation		
Inter-agency partnerships		

published its first national cybersecurity strategy only in December 2016 as the GCI report was being finalized. But China still does not have a developed set of cybersecurity metrics and its official statements continue to lack a strong sense of how to bring nuance to the claims of dire threat or degrees of cyber harm.

7.3   Other Chinese Assessments

The government's most comprehensive assessment of threat and the country's cybersecurity capability is implied more than expressed in its 2016 Cybersecurity Strategy, discussed in Chap. 1. The discussion here calls out what the strategy says about the country's capacities in cybersecurity. Above all, the strategy sees cyberspace offering both opportunity and threat to China: the chance to use advanced ICT for better cybersecurity as well a range of threats for which the country's network defences are patently inadequate.

The evolution in China's assessments of its own cybersecurity can be traced through its official surveys but these are fragmentary, often quite rudimentary and far from comprehensive. In the 2001 survey, only 25% of users had confidence in the security of online activity. In half-yearly surveys between 1999 and 2002, more than 30% of business users saw lack of information security as the main obstacle to online business (CNNIC 2002a: Sect. 7.3). By 2002, that had dropped to around 20% (CNNIC 2002b: 17). It was not until 2003 that CNNIC began asking how often users scanned their systems for viruses (CNNIC 2003: 26) and other security related practices. By 2005, the survey was asking about privacy actions by users, such as modification of default settings (CNNIC 2005: 16).

By 2017, the Chinese government was offering more structured and consistent reporting. For example, CNCERT/CC (2017: 14–20) offered the following positive trends through 2016:

- cybersecurity situation is generally stable
- no significant impact on normal operations
- cyberspace law and governance are clearer
- domain name system security is in good condition, and anti-attack ability increased significantly
- new security technology based on artificial intelligence is in full swing.

On the threat side, the list was somewhat longer:

- volume of mobile malware continued to rise rapidly and to have a significant impact on profits
- attacks from foreign sources were frequent
- large-scale distributed denial of service attacks, particularly through networked intelligent devices
- serious damage to data
- personal information disclosure, and 'derivative disaster' arising from that
- rampant fraud
- government-launched advanced persistent threats (APT) directly threatening national security and stability are becoming normal and are a particularly serious threat
- the number of network security attacks against industrial control systems is increasing, with many important safety incidents
- the dark web business model has matured
- ransomware and extortion software are proliferating
- the use of Internet of things intelligent device network attacks will continue to increase
- the security threat posed by the integration of the Internet and traditional industries is more complex
- the national origin of attacks continues to become more important.

According to the editors of China's 2016 Bluebook on cyberspace security, the country's situation in the field was "still grim and new manifestations of risks appeared" in the preceding year in spite of the great importance attached to security by the government and firms (Hui and Tan 2016: 8). They talked of "high-risk security vulnerabilities in parts of state-level information systems" and all too frequent security incidents in large e-commerce platforms. They described these as new features in part because of new threats. For example, they cited the first case where a foreign state's attack on China had been revealed, involving a hacker group dubbed "Ocean Lotus". Qihoo 360 uncovered the attacks in May 2015, naming China's maritime administration departments as a target since April 2014. As an example of the second trend, the attack on e-commerce platforms, they cited a May 2015 incident which paralysed Alipay, the country's largest such platform, and an attack on Ctrip, China's largest online travel business. Reports of the latter attack caused the company's share price on Nasdaq to fall by 20% at the opening of trade later that day (He 2015).

These analysts made plain their view that China could not escape changes in the global cyber ecosystem, to which Chinese criminals were in fact contributing. They mentioned XcodeGhost malware, using a new approach based on modification of the Xcode compiler. The attack infected over 3000 apps available in China's Apple App Store (including WeChat, Netease Cloud Music, AutoNavi, Didi Chuxing, Railway 12,306, and some banking apps). The authors cited the case of Chinese hackers whom police arrested in late 2015 for using a new technique called "brushing" in China. Reuters reported several months later (Reuters 2016), when the news of the attack broke in the Shanghai media, the share price on Alibaba's US listed entity fell 3.7% (The criminals first obtained 99 million sets of usernames and passwords from various websites, and then rented Alibaba's cloud computing service to use compromised accounts to place fake orders on e-commerce platform, Taobao, where they found a 20% match of accounts-some 20 million.).

7.4 International Assessments

Bearing out the amber assessment of the ITU GCI for China in the area of cyber crime, Europol (2016: 60) assessed that "China has an extensive and increasingly innovative digital underground". Of note, the report observed that the crime-related products and services available in China "mirrors that of Western underground markets", even if there is less use of traditional cybercrime forums in favour of reliance on "instant messaging or spam on existing (unrelated) fora to drum up business". China was seen as a "key source for tools and equipment relating to card crime, such as ATM and POS skimmers". The country was also called out for having "some of the highest global malware infection rates and consequently highest volumes of global Bots". China remained a leading sources of spam and was responsible for more than one quarter of worldwide DDoS attacks. China and Hong Kong (alongside India) were identified by a large number of European countries as having significant cyber criminal infrastructure.

Symantec (2017) paints a picture of China as amongst the top ten countries in the world for cyber insecurity, whether that is in terms of attacks generated from China or attacks on it. For example, China is ranked 6th in the number of countries with recorded cases of identity theft for the number of identities compromised (some 11 million): after the United States (791 million), France, Russia, Canada and Taiwan, and just ahead of South Korea and Japan (50).

PWC's global economic crime survey in 2016 reported that only 13% of China's economic fraud cases reported by respondents in the past 24 months came from cyber crime, compared with more than 50% in Hong Kong and Macau, and more than

25% in the Asia Pacific as a whole (PWC 2016: 3). This could reflect lower levels of monitoring and understanding, though the rate reported for China in the same survey in 2014 was 22%. The report suggested that Chinese mainland companies may not be "on top of the situation" and recommended that they ask themselves "whether they need to be doing more to secure themselves against the risk of money laundering, social engineering, and cybercrime" (12). McAfee reported in 2014, with high confidence in its estimate, that in China the net cost of cyber crime was equivalent to 0.63% of GDP, one of the highest figures for the G20 (McAfee 2014: 21).

7.5   Nine Strategic Tasks: How Is China Faring

Most approaches to grading national cybersecurity on a comparative basis rely on a standard set of indicators, mostly inputs (such as this or that government policy document, such as a policy for countering cyber crime). That is a defensible approach but it is shaped in part by the lack of data for outputs (such as the number of cybercrime convictions or the number of APTs successfully evicted from systems). A more appropriate means of assessing countries' comparative efforts would be one that also takes account of outputs, and probably also the specific circumstances of the country. But above all, an assessment of a country's performance cannot only be oriented toward the goals and objectives of its government. Different stakeholder groups will assess these differently, and their interests, alongside government interests, need to be factored in.

Table 7.2 offers a provisional and subjective interpretation by the author of how China's government is performing by mid-2017 in terms of the nine strategic tasks that it set itself. Each task is assessed under five headings that correspond to the five pillars of the Global Cybersecurity Agenda. The scale used is 1–5, where the lowest score (1) represents no landmark success, (3) represents moderate success, and (5) represents best achievable outcome with appropriate resources and commitment. This assessment framework differs from the ITU GCI in that the former assesses results and outputs against specific national objectives whereas the GCI only assesses inputs.

It should be noted that China performs poorly (below satisfactory) against all of the items in the Austin/Slay checklist introduced at the start of this chapter, except perhaps for item 7: "effective response capabilities for social threats (crimes) against individuals, including children and other vulnerable groups". China may be one of the more effective states in the world against this criterion, based in part on its intrusive social surveillance systems.

Many aspects of these assessments have been elaborated in earlier chapters. One of the most convincing sets of evidence for the low scores on most fronts is the high intensity and wide scope of the country's policy making in various fields of cybersecurity beginning with Xi's cyber power announcement of February 2014. Statements by Chinese leaders on the gravity of the challenges faced by the country provide additional evidence. Analysts do need to discount Chinese official statements

Table 7.2 Grading China's cybersecurity: its own tasks

	TECH	CAPAC	LEGAL	ORG'L	COOP'N
1. Defend cyberspace sovereignty	3	3	2	2	2
2. Uphold national security	4	2	5	5	2
3. Protect critical information infrastructure (CII)	1	1	1	1	1
4. Strengthening online culture	3	3	5	4	3
5. Combat cyber terror and crime	3	2	3	3	2
6. Improve cyber governance	3	2	2	2	3
7. Reinforce the foundations of cybersecurity	2	2	2	2	2
8. Enhance cyberspace defence capabilities	1	1	1	1	1
9. Strengthen international cooperation	3	2	2	3	3

for various propaganda and mobilization agendas, but even so, there is sufficient detail in the factual record to support the general thrust of the assessments above.

Moreover, the ITU GCI gives China modest scores on input-related assessments. For a country like China, with historic and chronic challenges of implementation at the grass roots of polices set at the top in Beijing, its scores on results (outputs) would of necessity be lower than its scores based on inputs. In almost all countries, compliance with policy, and even with law, is consistently weaker than the statements of intent. For cyberspace affairs, where visibility on many areas of activities for supervising authorities depends on technical monitoring of code and connections, where forensics can be quite complex, and where there are usually no witnesses subject to open investigation, compliance is a very big challenge from the outset regardless of the country.

China has not been able to consistently articulate the different domains of cyberse-curity. Though it does quite well on cyber crime, harassment and bullying, it leaves almost untouched in the public domain serious questions of critical infrastructure resilience and warfare. It has a mixed record in its articulation of cyber espionage threats, preferring to paint these very much in a "China versus America" lens, while ignoring (in public at least) threats from Russia, Israel, and even North Korea. China ignores most of the human and social aspects of cybersecurity, and has moved only since 2014 to address many of the legal questions. It is slowly improving its public policy on how different sections of the society must bear differentiated responsibil-ities. It has improved its articulation of the threat environment, including through better consistency and comprehensive reporting, but has dome less well in framing variegated response options. Beginning in 2014, China started to move toward a comprehensive suite of governmental, cross-sector, private-public, professional and civic organizations active in cybersecurity, and this intensified in 2016.

Like most countries, China has quite some difficulty in framing a national consensus on where to draw the line between sovereign capabilities and the global communities of practice (including R&D).

China has set in place effective monitoring of business threats and some rapid response capabilities at the enterprise level, including large corporations and SMEs. But it has not expanded this to study of the wider economic impacts of cyber threats and responses.

China is least developed in terms of nation-wide preparedness for the unlikely but credible threat of an extreme cyber emergency affecting the civil economy (as opposed to one or two sectors) or threatening national security interests (including international aspects).

References

Austin G, Slay J (2016) Australia's response to advanced technology threats: an agenda for the next government. UNSW Canberra, Canberra, Australian Centre for Cyber Security Discussion Paper#3

CNCERT/CC (2016) A summary of China's internet security situation in 2015. National Computer Network Emergency Technology Processing Coordination Center, April 2016. http://www.cac.gov.cn/files/pdf/wlaq/Annual%20Report/CNCERT2015.pdf (In Chinese)

CNCERT/CC (2017) A summary of China's internet security situation in 2016. National Computer Network Emergency Technology Processing Coordination Center, April 2017. http://www.cac.gov.cn/wxb_pdf/CNCERT2017/2016situation.pdf (In Chinese)

CNNIC (2002a) 9th statistical survey of the development of the internet in China. China Internet Network Information Center, January. http://www.cnnic.net.cn/download/manual/en-reports/9.pdf

CNNIC (2002b) 10th statistical survey of the development of the internet in China. China Internet Network Information Center, July. http://www.cnnic.net.cn/download/manual/en-reports/10.pdf

CNNIC (2005) 16th statistical survey report on the internet development in China. China Internet Network Information Center, July (URL no longer available)

Europol (2016) IOCTA 2016 Internet organised crime threat assessment. https://www.europol.europa.eu/activities-services/main-reports/internet-organs-crime-threat-assessment-iocta-2016

Feihua News (2017) Experts expose ransomware's true face. What measures did relevant departments in our country take? 15 May 2017. https://news.fh21.com.cn/rdsj/4912207_6.html (In Chinese)

He W (2015) Shares in Chinese online travel agent Ctrip plunge after hack attack stops service. South China Morning Post, 28 May 2015. http://www.scmp.com/tech/enterprises/article/1811165/chinas-largest-online-travel-agent-ctrip-taken-offline-hackers

Hui Z, Tan Q (2016) Cyberspace security in the era of data economy: global and Chinese contexts. In: Hui Z, Tan Q (eds) Annual report on development of cyberspace security in China. Social Sciences Academy Press, Blue Book, Beijing (In Chinese). 1–16

ITU (2007) ITU Global Cybersecurity Agenda (GCA) High-Level Experts Group (HLEG). Report of the Chairman to HLEG. International Telecommunication Union, Geneva, Switzerland. https://www.itu.int/en/action/cybersecurity/Documents/gca-chairman-report.pdf

ITU (2014) Global 2014 results. International Telecommunication Union. Geneva, Switzerland. https://www.itu.int/en/ITU-D/Cybersecurity/Documents/GCI_Global_2014_results.pdf

ITU (2015) Global cybersecurity index and wellness profiles. International telecommunication Union, Geneva, Switzerland. https://www.itu.int/dms_pub/itu-d/opb/str/D-STR-SECU-2015-PDF-E.pdf

ITU (2017) Global cybersecurity index 2017. International telecommunication Union, Geneva, Switzerland, 19 July 2017. https://www.itu.int/dms_pub/itu-d/opb/str/D-STR-GCI.01-2017-R1-PDF-E.pdf

Marro N (2016) The five levels of cyber security in China. China Business Review, 5 Dec 2016. http://www.chinabusinessreview.com/the-5-levels-of-information-security-in-china/

McAfee (2014) Net losses: estimating the global cost of cybercrime. Economic impact of cybercrime II. Cent for Strateg Int Stud. https://www.mcafee.com/de/resources/reports/rp-economic-impact-cybercrime2.pdf

PWC (2016) PwC global economic crime survey 2016—mainland China, Hong Kong and Macau Supplement. https://www.pwccn.com/en/migration/pdf/forensic-economic-crime-survey-2016.pdf

Qihoo 360 (2017a) 2016 China internet security report, 12 Feb 2017. http://zt.360.cn/1101061855.php?dtid=1101062514&did=490278985 (In Chinese)

Qihoo 360 (2017b) Research on internet fraud trends for 2016, 13 Jan 2017. https://e70efc.lt.yunpan.cn/lk/cWHqQF5BHYVZF (In Chinese)

Qu W (2010) China's path to informatization, English ed. Cengage Learning Asia

Reuters (2016) Hackers attack 20 million accounts on Alibaba's Taobao shopping site, 4 Feb 2016. http://www.reuters.com/article/us-alibaba-cyber-idUSKCN0VD14X

Symantec (2017) Internet security threat report, vol 22. https://www.symantec.com/content/dam/symantec/docs/reports/istr-22-2017-en.pdf

www.gov.cn (2007) Ministry of public security and others issued notice on "information security level protection management approach", 24 July 2007. http://www.gov.cn/gzdt/2007-07/24/content_694380.htm (In Chinese)

Xi Jinping (2016) Speech at the work conference for cybersecurity and informatization, Posted on 19 April 2016, updated on 26 April 2016. https://chinacopyrightandmedia.wordpress.com/2016/04/19/speech-at-the-work-conference-for-cybersecurity-and-informatization/

Chapter 8
The Next Wave

Abstract The short conclusion highlights key findings of the book, emphasizing the need to look beyond a monochromatic characterization of national cybersecurity to one based on diverse mission sets and stakeholder sets. It comments on the balance between technology, society and international political economy in shaping the future of cybersecurity in China.

8.1 China as a Case Study in National Cyber Security

This book suggests that a country as such does not have and cannot have "cybersecurity" as if that were a quality or commodity. One can only speak meaningfully of which actors in a country enjoy cyber security, at what levels, and in what circumstances. Cybersecurity is a complex socio-technical system that combines a sense of threat (insecurity) and a sense of confidence in the quality of defence (security) of different actors. It is also in part a technical reality: where a victim can feel secure even while foreign states, criminals, corporations, or even his/her own government rampage through his/her cyber space in abusive and exploitative ways. China is a special case for the relationships and contrasts between the cybersecurity interests of its citizens, its corporations and its government. The Chinese government is probably one of the most highly targeted in the world by cyber espionage of other states, and by its own agencies.

Perceptions of cybersecurity are social and psychological phenomena that might usefully be measured in assessing national cybersecurity but that are rarely taken into account. In the case of China, there has been very little scholarly work on the cybersecurity perceptions of different stakeholders, or the organizational cultures in which they must try to establish their own cyber security. Some basic surveys have been in place for almost two decades but the political and institutional constraints on meaningful and penetrating analysis of the phenomena have been very powerful. Universities and research centres have until recently, with few exceptions, virtually ignored the non-technical dimensions of security in cyber space except to reinforce government policy interests.

© The Author(s) 2018

G. Austin, *Cybersecurity in China*, SpringerBriefs in Cybersecurity,
https://doi.org/10.1007/978-3-319-68436-9_8

Cybersecurity in one country or another has to be evaluated from a multi-layered point of view that can only be viewed as an interactive, large and highly diverse matrix with dynamic trend aspects. A static snapshot list cannot give an adequate accounting. The number and type of elements in the matrix must be far wider than in most lists used in compiling indexes of national cybersecurity capability.

What does this mean for China's ambition announced in 2014 of becoming a cyber power? It is reasonable to frame an ambition, as China's leaders have, to become a world leader in cybersecurity technologies, especially their potential military application. However, national competence in designing and developing cyber security technologies is an activity of the cyber industrial complex (enterprises, researchers and investors). It is not an activity of the broader society or most of its key institutions. To become a cyber power on the global stage, China will need to develop the work force that can support that; while simultaneously crafting the international alliances that will underpin the domestic accumulation of attributes of cyber power. Even for the more limited ambition of becoming the most powerful cyber-enabled police state in the world through an Orwellian system of social credits, China faces huge social and institutional obstacles. How will it train its two million public security police to protect systems and secrets in cyber space, while simultaneously training them to act on or evaluate the social credit of every single Chinese adult citizen.

China's ambitions in cyber space against its own citizens are seen by liberal democracies as an existential threat and therefore China will continue to face opposition in many of its cybersecurity policies from them. Information wars are intensifying and this will result in increasing attacks on the credibility and security of Chinese cybersecurity systems from outside the country.

8.2 Sub-national Cybersecurity, Actors and Mission Sets

Having rejected the value of a "national cybersecurity" assessment, the book provides some leads to how analysts might usefully assess sub-national aspects of the phenomenon by reference to specific mission sets (and by implication, to various stakeholders with some divergent interests). These mission sets might include, but not be limited to, countering cybercrime, protecting consumers online, defence of CII, or censorship practices. Actors in these mission sets in China demonstrate vastly different capabilities in cybersecurity technologies and management techniques. China's heavy emphasis on online monitoring has underpinned its strengths in censorship, and has probably had payoffs in countering cyber crime, simply because there is more surveillance.

While China's cybersecurity industry is going from strength to strength, it is not a world leader in many fields, with several exceptions possibly being in malware detection capability, in some aspects of quantum computing, and in applications of artificial intelligence to domestic surveillance. The country's universities have only recently opened necessary pathways for greater and more comprehensive innovation and understanding of cybersecurity, and these pathways may take a decade or more to mature.

8.3 Artificial Intelligence

Advanced artificial intelligence will dominate the next wave of cyber security in China. This is in part a technological inevitability, but it is also driven by the demands of an ever expanding police state intent on controlling the political orientation of its citizens. But in China's case, advanced AI offers the only solution to the estimated skills deficit of 1.4 million cyber security professionals by 2020. China will need foreign specialists and corporations to achieve its ambitions in this field of research and industrial development. As of 2018, several leading foreign firms in China are serving as the henchmen of 'big brother'.

8.4 Quantum Computing

China has, without any doubt, made great strides and world class achievements in quantum communications. At the same time, it makes consistent propaganda about an alleged first-rank position in related research and applications and how that will help cement the country's cybersecurity future. The hope may be somewhat misplaced. Even though leading governments elsewhere stake much on quantum computing as a security game-changer, the potential of the technologies as a widely deployable and usable form of advanced cybersecurity is still untested. China will do better if it looks to socio-technical and management approaches for more highly secure computing than to miracle technologies.

8.5 Collaborate to Compete

The majority view of the Chinese leadership is that key agencies in the government and leading corporations cannot have cybersecurity in China unless the country pursues collaborative policies with other major powers and world leading foreign firms. As the leaders promote indigenization of the cybersecurity industry, and as that process takes greater hold, the global scene will continue to race ahead of China's domestic capacities, and Chinese specialists working abroad will become increasingly active in that technological and commercial advance.

8.6   Complexity

China's leaders can, of course, handle complexity in certain fields of activity, including the preparation of high-tech armed forces for use on land, in the air, in or under the sea, and in outer space. The challenges of complexity represented by the strategic tasks they have set themselves in cyber space are however on another level than that, including military uses of cyberspace and protection of national CII. The country's leading specialists know what they are dealing with but their leaders don't want to admit that the character of the information age might simply be incompatible with China's system of government, political priorities and overall stage of informatization.

8.7   Social Resistance and Modernization

Cyberspace is a contested domain. The Chinese government's national cyber security strategy of December 2016 makes plain that the leaders see themselves caught up in a political battle and a modernization whirlwind the likes of which few of them imagined ten years ago. Establishing a Communist Chinese version of security in this environment may well remain an impossible dream, if only because the actors most capable of, and intent on spoiling, the authoritarian dream are part of the most powerful and technologically most advanced alliance ever seen in human history. The struggle is only just beginning, and China's adversaries are only just starting to mobilise their political, economic, scientific, social and military powers to ensure that the new authoritarian view of cyberspace in China is defeated. The outcome of the battle will be determined by the character and capabilities of individual combatants on each side, especially by Chinese citizens at large, but also by the deepening of globalization and the imperative of "collaborate to compete". China's current campaign of security in cyberspace through national virtue, patriotism and hyper-vigilance against alternative thinking will see more great victories for the government and many defeats for Chinese citizens. But at the end of the day, there may be no "great firewall" big enough or long enough to fence China off from the information revolution. This book demonstrates the need for much deeper and more comprehensive research on the cyber power of the Chinese state.

Made in United States
North Haven, CT
29 August 2023

40903397R00085